Time Travel and Warp Drives

Time Travel

and Warp Drives

A Scientific Guide

to Shortcuts

through Time and Space

Allen Everett and Thomas Roman

The University of Chicago Press

Chicago and London

ALLEN EVERETT is professor emeritus of physics at Tufts University.
TOM ROMAN is a professor in the Mathematical Sciences Department at Central Connecticut
State University. Both have taught undergraduate courses in time-travel physics.

The University of Chicago Press, Chicago 60637
The University of Chicago Press, Ltd., London
© 2012 by The University of Chicago
All rights reserved. Published 2012.
Printed in the United States of America

21 20 19 18 17 16 15 14 13 12   1 2 3 4 5

ISBN-13: 978-0-226-22498-5 (cloth)
ISBN-10: 0-226-22498-8 (cloth)

Library of Congress cataloging-in-Publication Data

Everett, Allen.
    Time travel and warp drives : a scientific guide to shortcuts
through time and space / Allen Everett and Thomas Roman.
        p. cm.
    Includes bibliographical references and index.
    ISBN-13: 978-0-226-22498-5 (cloth : alk. paper)
    ISBN-10: 0-226-22498-8 (cloth : alk. paper)
    1. Time travel.   2. Space and time.   I. Roman, Thomas.   II. Title.
QC173.59.S65E94   2012
530.11—dc23

                                                            2011025250

⊗ This paper meets the requirements of
ANSI/NISO Z39.48–1992 (Permanence of Paper).

To my loving wife, Cecilia,
and to my parents (T. R.)

In memory of my late beloved wife and cherished
best friend, Marylee Sticklin Everett. For more
than 42 years of love, companionship, support,
and wonderful memories, thank you. (A. E.)

Contents

Preface > ix

Acknowledgments > xi

1   Introduction > 1

2   Time, Clocks, and Reference Frames > 10

3   Lorentz Transformations and Special Relativity > 22

4   The Light Cone > 42

5   Forward Time Travel and the Twin "Paradox" > 49

6   "Forward, into the Past" > 62

7   The Arrow of Time > 76

8   General Relativity: Curved Space and Warped Time > 89

9   Wormholes and Warp Bubbles: Beating the
Light Barrier and Possible Time Machines > 112

10   Banana Peels and Parallel Worlds > 136

11   "Don't Be So Negative": Exotic Matter > 158

12   "To Boldly Go . . ."? > 181

13   Cylinders and Strings > 196

14   Epilogue > 218

Appendix 1. Derivation of the Galilean Velocity Transformations > 225

Appendix 2. Derivation of the Lorentz Transformations > 227

Appendix 3. Proof of the Invariance of the Spacetime Interval > 232

Appendix 4. Argument to Show the Orientation of the x',t' Axes
Relative to the x,t Axes > 234

Appendix 5. Time Dilation via Light Clocks > 236

Appendix 6. Hawking's Theorem > 241

Appendix 7. Light Pipe in the Mallett Time Machine > 250

Bibliography > 253

Index > 259

Preface

In part, our motivation for writing this book is the classes that we have taught on the subject at our respective universities, Tufts (A. E.) and Central Connecticut State (T. R.). Many, but not all, of our students were science fiction buffs. They ranged from present or prospective physics majors to fine arts majors; several of the latter did very well and were among the most fun to teach. The courses afforded us an opportunity, unusual for theoretical physicists, to give undergraduates some access to our own research, using essentially no mathematics beyond high school algebra. We are grateful to all of the students in those classes over the years for their enthusiasm and intellectual stimulation.

Our aim here was to write a book for people with different levels of math and physics backgrounds, skills, and interests. Since we believe that what currently is on offer is either too watered down or too sensationalistic, we decided to try our hand. The level of this book is intended for a person who is perhaps a *Star Trek* fan or who likes to read *Scientific American* occasionally, but who finds it not detailed enough for a good understanding of the subject matter. We assume that our reader knows high school algebra, but no knowledge of higher mathematics is assumed. A basic physics course, although helpful, is not necessary for understanding. However, the reader will need to expend some intellectual effort in grappling with the concepts to come. We realize that not every reader will be interested in the same level of detail. Therefore many (although not all!) of the mathematical details have been placed in appendixes, for those who are interested in more "meat." Our feeling is that even readers who want to "skip the math" will still find plenty of topics to interest them in our book. So, although we do not expect *every* reader to understand every single item in the book, we have aimed to provide a stimulating experience for *all* readers. Interactive Quicktime demonstrations that illustrate some of the concepts in the book can be found at http://press.uchicago.edu/sites/timewarp/.

< ix >

Acknowledgements

We would like to thank Chris Fewster, Larry Ford, David Garfinkle, Jim Hartle, Bernard Kay, Ken Olum, Amos Ori, David Toomey, Doug Urban, and Alex Vilenkin for useful discussions. We would also like to thank Dave LaPierre and Tim Ouellette for reading the manuscript and providing us with critical comments. Special thanks to Tim Ouellette for applying his considerable editing skills to the manuscript and for his help with the figures. Our initial editor at the University of Chicago Press, Jennifer Howard, gave us constant enthusiastic support during the early stages of this work. Finally, we wish to thank our present editors, Christie Henry, Abby Collier, and especially Mary Gehl, for all their help in turning this manuscript into an actual book.

Allen would like to thank his former student, and later colleague, Adel Antippa, for dragging him in 1970 into what proved to be a stimulating collaborative study of the possible physics of tachyons. Adel's student, now Professor Louis Marchldon, also made important contributions to this work. This laid a foundation for Allen's renewed interest a quarter of a century later in the physics of superluminal travel and time machines, when interesting new developments began to occur. Allen would also like to extend a special acknowledgment to Mrs. Gayle Grant, the secretary of the Physics and Astronomy Department at Tufts. Over a number of years, Gayle's efficiency and dependability have contributed in countless ways to all aspects of Allen's professional career, including those connected with this book. Perhaps even more important, her unfailing cheerful friendliness, to faculty and students alike, was an important factor in making the Physics Department a very pleasant place to work.

Tom would like to thank the National Science Foundation for partial support under the grant PHY-0968805.

< xi >

1

Introduction

As humans, we have always been beck-oned by faraway times and places. Ever since man realized what the stars were, we have wondered whether we would ever be able to travel to them. Such thoughts have provided fertile ground over the years for science fiction writers seeking interesting plotlines. But the vast distances separating astronomical objects forced authors to invent various imaginary devices that would allow their characters to travel at speeds greater than the speed of light. (The speed of light in empty space, generally denoted as c by physicists, is 186,000 miles/second.) To give you an idea of the enormous distances between the stars, let's start with a few facts. The nearest star, Proxima Centauri (in the Alpha Centauri star system) is about 4 light-years away. A light-year is the distance that light travels in a year, about 6 trillion miles. So the *nearest* star is about 24 trillion miles away. It would take a beam of light traveling 186,000 miles per second, or a radio message, which would travel at the same speed, 4 years to get there.

On an even greater scale, the distance across our Milky Way galaxy is approximately 100,000 light-years. Our nearby neighbor galaxy, Andromeda, is about 2,000,000 light-years away. With present technology, it would take some tens of thousands of years just to send a probe, traveling at a speed far less than c, to the nearest star. It's not surprising then that science fiction writers have long imagined some sort of "shortcut" between the stars involving travel faster than the speed of light. Otherwise it is difficult to see how one could have the kinds of "federations" or "galactic empires" that are so prominent in science fiction. Without shortcuts, the universe is a very big place.

And what about time, that most mysterious feature of the universe? Why is the past different from the future? Why can we remember the past and not the future? Is it possible that the past and future are "places" that can be visited, just like other regions of space? If so, how could we do it?

< 1 >

This book examines the possibility of time travel and of space travel at speeds exceeding the speed of light, in light of physics research conducted during the last twenty years or so. The ideas of faster-than-light travel and time travel have long existed in popular imagination. What you may not know is that some physicists study these concepts very seriously—not just as a "what might someday be possible" question, but also as a "what can we learn from such studies about basic physics" question.

Science fiction television and movie series, such as *Star Trek*, contain many fictional examples of faster-than-light travel. Captains Kirk or Picard give the helmsman of the starship *Enterprise* an order like, "All ahead warp factor 2." We're never told quite what that means, but we're clearly meant to understand that it means some speed greater than the speed of light (c). Some fans have speculated that it refers to a speed of 2^2c, or four times the speed of light. These speeds are supposed to be achieved by making use of the *Enterprise*'s "warp drive." This term was never explained and seems to be merely a nice example of the good "technobabble" usually necessary in a piece of science *fiction* to make things sound "scientific." But by chance—or good insight—*Star Trek*'s "warp drive" turns out to be an apt description of one conceivable mechanism for traveling at faster-than-light speed, as we shall discuss later in some detail. For this reason, we will use the term "warp drive" from now on to mean a capacity for faster-than-light travel.

By analogy with the term "supersonic" for speeds exceeding the speed of sound in air, speeds greater than the speed of light are often referred to in physics as "superluminal speeds." However, superluminal travel seems to involve a violation of the known laws of physics, in this case, Einstein's special theory of relativity. Special relativity has built into it the existence of a "light barrier." The terminology is intended to be reminiscent of the sound barrier encountered by aircraft when their speed reaches that of sound and which some, at one time, thought might prevent supersonic flight. But whereas it proved possible to overcome the sound barrier without violating any physical laws, special relativity seems to imply that superluminal travel, that is, an actual warp drive, is absolutely forbidden, no matter how powerful some future spaceship's engines might be.

Time travel also abounds in science fiction. For example, the characters in a story may find themselves traveling back to our time period and becoming involved with a NASA space launch on Earth, perhaps after passing through a "time gate." Often in science fiction, the occurrence of backward time travel seems to have nothing to do with the existence of a warp drive for spaceships;

the two phenomena of superluminal travel and time travel appear quite unrelated. In fact, we shall see that there is a direct connection between the two.

Science fiction writers often provide imaginative answers to questions beginning with the word "what."—"What technological developments might occur in the future?" —but in general, science fiction does not provide answers to the question of "how". It usually provides no practical guidance as to just how some particular technological advance might be achieved. Scientists and engineers by contrast work to answer "how," attempting to extend our knowledge of the laws of nature and to apply this knowledge creatively in new situations.

The fact that science, in due course, frequently has provided answers as to how some imagined technological advance can actually be achieved may tend to lead to an expectation that this will always occur. But this is not necessarily true. Well-established laws of physics often take the form of asserting that certain physical phenomena are absolutely forbidden. For example, as far as we know, no matter what occurs, the total amount of energy of all kinds in the universe does not change. That is, in the language of physics, energy is said to be "conserved," as you were probably told in your high school and university science courses.

Although works of science fiction usually cannot address the "how" questions, they often serve science through their explorations of "what." By envisioning conceivable phenomena outside of our everyday experience, they may offer science possible avenues of experimentation. Some of the chapters of this book contain suggested science fiction readings or films that relate to the subject matter of the chapter and can prove helpful in visualizing various scenarios which might occur if, for example, time travel became possible.

A writer of science fiction is at liberty to imagine a world in which humans have learned to create energy in unlimited quantities by means of some imaginary device. However, a physicist will say that, according to well-established physical laws, this will not be possible, no matter how clever future scientists and engineers may be. In other words, sometimes the answer to the question "How can such and such a thing be done?" is "In all probability, it can't." We must be prepared for the possibility that we will encounter such situations.

Unless we specify otherwise, the term "time travel" will normally mean time travel into the past, which is where the most interesting problems arise. As a convenient shorthand we will refer to a device that would allow this as a "time machine" and to a process of developing a capacity for backward time travel as "building a time machine." This implies the possibility that you could go back in time and meet a younger version of yourself. In physics jargon, such a circular path in space and time is referred to as a "closed timelike curve." It is closed because you can return to your starting point in both space and time. It is called "timelike" because the time changes from point to point along the curve. The statement that a closed timelike curve exists is just a fancy way of saying that you have a time machine.

It would seem that time travel into the past should also be impossible outside the world of science fiction simply on the basis of ordinary common sense because of the paradoxes to which it seems to lead. These are typified by what is often called the "grandfather paradox." According to this scenario, were it possible to travel into the past, a time traveler could in principle murder his own grandfather before the birth of his mother. In this case he would never be born, in which case he would never travel back in time to murder his grandfather, in which case he would be born and murder his grandfather, and so on and so on forever. In summary, the entrance of the grandson into the time machine prevents his entrance into the machine. Such paradoxical situations that involve logical contradictions are called "inconsistent causal loops." The laws of physics should allow one to predict that, in a given situation, a certain event either does or does not occur. Hence, they must be such that inconsistent causal loops are not allowed.

For some time, warp drives and time machines were generally believed to be confined to the realm of science fiction because of the special relativistic light barrier and the paradoxes involved with backward time travel. Over the past several decades, the possibility that superluminal travel and backward time travel might actually be possible, at least in principle, has become a subject of serious discussion among physicists. Much of this change is due to an article entitled "Wormholes, Time Machines, and the Weak Energy Condition," by three physicists at the California Institute of Technology: M. S. Morris, K. S. Thorne, and U. Yurtsever. Their article was published in 1988 in the prestigious journal *Physical Review Letters*. (You will learn something of the meaning of that strange-sounding phrase "weak energy condition" later.) The senior author, K. S. Thorne (who is the Feynman Professor of Theoretical Physics at Caltech), is one of the world's foremost experts on the general theory of relativity, which is Einstein's theory of gravity. The discovery of the latter theory followed that of special relativity by about a decade. General relativity offers potential loopholes that might allow a sufficiently advanced civilization to find a way around the light barrier.

As far as time travel into the future is concerned, it is well understood in physics—and has been for a good part of a century—that it is not only possible but also, indeed, rather commonplace. Here, by "time travel into the future," we implicitly mean at a rate greater than the normal pace of everyday life. Forward time travel is, in fact, directly relevant to observable physics, since it is seen to occur for subatomic particles at high energy accelerators, such as that at Fermi National Laboratory, or the new Large Hadron Collider (LHC) at the European Organization for Nuclear Research (CERN) in Geneva, where such particles attain speeds very close to the speed of light. (Sending larger masses, such as people or spaceships, a significant distance into the future, while possible in principle, requires amounts of energy which are at present prohibitively large.)

We begin the exploration of forward time travel with a brief discussion of the meaning of time itself in physics. We will then have to do some thinking about just what the phrase "time travel" means. For example, what would we expect to observe if we traveled in time, and what would non–time travelers around us see? Like a number of things in this book, answering these questions requires stretching the imagination to envision phenomena that you have never actually encountered or probably even thought carefully about.

After that, you will learn the fundamentals of Einstein's special theory of relativity. The discovery of special relativity is one of the great intellectual achievements in the history of physics, and yet the theory involves only rather simple ideas and no mathematics beyond high school algebra. Again, however, to understand what is going on you have to be prepared to stretch your thinking beyond what you observe in your everyday life. Special relativity describes the behavior of objects when their speed approaches the speed of light. As we will see, special relativity leaves no doubt that forward time travel is possible. We will discuss one of the most remarkable predictions of special relativity, namely, that a clock appears to run slower when it is moving relative to a stationary observer, an effect called "time dilation." This effect becomes significant when the speed of the clock approaches c. Time dilation is closely related to what is called the "twin paradox." This is essentially the same phenomenon that is responsible for the "forward time travel" seen to occur for elementary particles at Fermilab and the LHC.

At first glance, faster-than-light travel might seem to be a natural extension of ordinary travel at sub-light speeds, just requiring the development of much more powerful engines. Space travel in many science fiction stories of the 1930s and '40s involved no violations of fundamental laws of physics. The speculation of science fiction began to be realized in practice about a quarter of a century later, when Neil Armstrong took his "one small step" onto the surface of the moon. However, superluminal travel seems to involve a violation of the known laws of physics, in this case, the special theory of relativity, with its light barrier.

In the absence of a time machine, everyday observations tell us that the laws of physics are such that effects always follow causes in time. Thus the effect cannot turn around and prevent the cause, and no causal loop can occur. This is no longer true in the presence of a time machine, since then a time traveler can observe the effect and then travel back in time to block the cause. Therefore it would appear that the existence of time machines—that is, backward time travel—is forbidden just by common sense. Moreover, we will see that in special relativity, backward time travel becomes closely connected to superluminal travel, so that the same "common sense" objections can be raised to the possibility of a warp drive, in addition to the light barrier problem.

Einstein's theory of gravity, general relativity, introduces a new ingredient into the mix. It combines space and time into a common structure called "spacetime." Space and time can be dynamical—spacetime has a structure that can curve and warp. Einstein showed that the warping of the geometry of space and time due to matter and energy is responsible for what we perceive as gravity. We will introduce you to some of the ideas of general relativity and its implications. One consequence that we will discuss is the black hole, which is believed to be the ultimate fate of the most massive stars. When such a star dies, it implodes on itself to the point where light emitted from the star is pulled right back in, rendering the object invisible. We will point out that sitting next to (or orbiting) a black hole also affords a possible means of forward time travel that is different from the time dilation of moving clocks discussed earlier.

As we will find, the laws of general relativity at least suggest that it is possible to curve, or warp, space in such a way as to produce a shortcut through space, and perhaps even time, which is known to general relativists as a "wormhole." Wormholes are one of the staple features of several science fiction series: *Star Trek Deep Space Nine*, *Farscape*, *Stargate SG1*, and *Sliders*. Several years after the article by Morris, Thorne, and Yurtsever, a possibility for actually constructing a warp drive was presented in a 1994 article by Miguel Alcubierre, then at the University of Cardiff in the United Kingdom, which was published in the journal *Classical and Quantum Gravity*. By making use of general relativity, Alcubierre exhibited a way in which empty spacetime could be curved, or warped, in such a way as to contain a "bubble" moving at an arbitrarily high speed as seen from outside the bubble. One might call such a thing a "warp bubble." If one could find a way of enclosing a spaceship in such a bubble, the spaceship would move at superluminal speed, for example, as seen from a planet outside the bubble, thus achieving an actual realization of a "warp drive." Another kind of warp drive was suggested by Serguei Krasnikov at the Central Astronomical Observatory in St. Petersberg, Russia in 1997. This "Krasnikov tube" is effectively a tube of distorted spacetime that connects the earth to, say, a distant star. From what we have said before about the connection between superluminal travel and backward time travel, one would expect that wormholes and warp bubbles could be used to construct time machines. This is indeed the case, as we will also show.

What is known about how one might actually build a wormhole or a warp bubble? We'll see that, while not hopeless, the prospect doesn't appear very promising. One disadvantage they all share is that they require a most unusual form of matter and energy, called "exotic matter," or, "negative energy." (In view of Einstein's famous equivalence relation between mass and energy, $E = mc^2$, we will frequently use the two terms "mass" and "energy" interchangeably.) A theorem by Stephen Hawking (the former Lucasian Professor of Mathematics at Cambridge University, the same chair once held by Isaac Newton) shows that, loosely speaking, if you want to build a time machine in a finite region of time and space, the presence of some exotic matter is required. As it turns out, the laws of physics actually allow the existence of exotic matter or negative energy. However, those same laws also appear to place severe restrictions on what you can do with it. Over the last fifteen years, there has been a great deal of work, much of it by Larry Ford of Tufts University and one of the authors (Tom), on the question of what restrictions, if any, the laws of physics impose on negative energy. We will describe some of what has been learned and its implications for the likelihood of constructing wormholes and warp drives.

One might well think that the potential paradoxes, such as the grandfather paradox, make it pointless to even consider the possibility of backward time travel. However, as we'll see, there are two general approaches that could allow the laws of physics to be consistent even if backward time travel is possible. Each of these is illustrated in numerous works of science fiction, but one or the other must turn out to have a basis in the actual laws of physics, if those laws allow one to build a time machine.

The first possibility is that it could be that the laws of physics are such that whenever you go to pull the trigger to kill your grandfather something happens to prevent it—you slip on a banana peel, for example (we like to call this the "banana peel mechanism"). This theory is, logically, perfectly consistent.

It is somewhat unappealing, however, because it's a little hard to understand how the laws of physics can always arrange to ensure the presence of a suitable banana peel.

The other approach makes use of the idea of parallel worlds. According to this idea, there are two different worlds: in one you are born and enter the time machine, and in the other you emerge from the time machine and kill your grandfather. There is no logical contradiction in the fact that you simultaneously kill and do not kill your grandfather, because the two mutually exclusive events happen in different worlds. Surprisingly there is an intellectually respectable idea in physics called the "many worlds interpretation of quantum mechanics," first introduced in an article in *Reviews of Modern Physics* way back in 1957 by Hugh Everett (no relation to Allen as far as we know). According to (the other) Everett there are not just two parallel worlds but infinitely many of them, which, moreover, multiply continuously like rabbits.

In a 1991 *Physical Review* article, David Deutsch of Oxford University (one of the founders of the theory of quantum computing) pointed out that if the many worlds interpretation is correct (and Professor Deutsch is convinced that it is), it is possible that a potential assassin, upon traveling back in time, would discover that he had also arrived in a different "world" so that no paradox would arise when he carried out the dastardly deed. Allen analyzed this idea in somewhat greater detail in a 2004 article in the same journal. He found that the many worlds interpretation, if correct, would indeed eliminate the paradox problem—but at the cost of introducing a substantial new difficulty, which we'll explain later.

Many physicists find the ideas involved in either approach to the solution of the paradox problem so distasteful that they believe, or at least certainly hope, that the laws of physics prohibit the construction of time machines. This is a hypothesis that Stephen Hawking has termed the "chronology protection conjecture." While this conjecture may very well prove to be correct, at the moment it remains only a conjecture, essentially an educated guess that has not been proved. We'll discuss some of the evidence for and against the conjecture.

Another set of situations in which backward time travel can occur involves the presence of one of several kinds of infinitely long, string-like or rotating cylindrical systems. In each of these cases it is possible, by running in the proper direction around a circular path enclosing the object in question, to return to your starting point in space before you left.

One model of the rotating cylinder type, due to Professor Ronald Mallett of the University of Connecticut, has received considerable attention lately in several places, including an article in the physics literature and Mallett's book, *Time Traveler* (2006). Mallett suggested that a cylinder of laser light, carried perhaps by a helical configuration of light pipes, could be used as the basis of a time machine. Two published articles, one by Ken Olum of Tufts and Allen and another by Olum alone, definitively showed that the Mallett model has serious defects, which we will discuss.

Finally, we will summarize where the subject stands today and what the prospects are for the future. How trustworthy can our conclusions be, given the present state of knowledge? How can we predict what twenty-third-century technology will be like, given twenty-first-century laws of physics? Might not future theories overturn these ideas, as so often has happened in the history of science? We give some partial answers to these questions.

2
Time, Clocks, and Reference Frames

As happens sometimes, a moment
settled and hovered and remained for
much more than a moment. And sound
stopped and movement stopped for
much, much more than a moment.
Then gradually time awakened again
and moved sluggishly on.

JOHN STEINBECK, *Of Mice and Men*

These lines from Steinbeck's novel capture the experience we have all had of the varying flow of personal time. Our subjective experience of time can be affected by many things: catching the fly ball that wins the game, winning the race, illness, drugs, or a traumatic experience. It is well known that drugs, such as marijuana and LSD, can change—sometimes profoundly in the latter case—the human perception of time. People who have been in car crashes report the feeling of time slowing down, with seconds seeming like minutes. The windshield appears to crack in slow motion due to the trauma of the accident. If our subjective experience of time is so fluid, we might ask, "Well then, what is time . . . *really*?" Most of us can give no better answer than Saint Augustine in the *Confessions*: "What then is time? If no one asks of me, I know; if I wish to explain to him who asks, I know not." Augustine's answer somewhat anticipates Supreme Court justice Potter Stewart's well-known definition of obscenity, delivered from the bench: "I know it when I see it."

In this book we are concerned with measures of time that do not depend on the variations and vagaries of human perception. Physicists do not at all discount the importance of the problem of the human cognition of time, but it is,

< 10 >

at present, too difficult a problem for us to solve. Instead our emphasis will be on what modern physics has learned about the subject of time. In our (admittedly biased) opinion, the most valuable insights we have about the nature of time are due to advances in physics. The description, at least in part, of what we have learned over the years of the twentieth and early twenty-first centuries form much of the core of this book. Hopefully you will find these revelations as fascinating as we do. However, before we embark on this journey, let us first pay a brief visit to a comfortable nineteenth-century living room, where a discussion is happening in front of a warm fireplace . . . .

Time Travel à la Wells

"The Time Traveller (for so it will be convenient to speak of him) was expounding a recondite matter to us. His grey eyes shone and twinkled, and his usually pale face was flushed and animated." So opens the most famous time travel story in literature, H. G. Wells's *The Time Machine*. The Time Traveller claims to his dinner guests that "Scientific people know very well that Time is only a kind of Space." The guests understandably protest that, although we are free to move about in the three dimensions of space, we do not have the same freedom to move around in time. The Time Traveller then shows them a model of a machine that, he claims, can travel in time as easily as we travel through space. He turns the machine on and it spins around, becomes indistinct, and promptly vanishes. The guests then discuss what has become of the machine and whether it has traveled into the past or the future.

One guest argues that it must have gone into the past, because if it went into the future it would still be visible on the table, having had to travel through the intervening times between its starting time and the present moment. Another guest counters that if the machine went into the past, then it would have been visible when they first came into the room during this and previous dinner visits. The Time Traveller goes on to explain that the machine is invisible to them because it is traveling through time at a much greater rate than they are. As a result, by the time they "get to" some moment, the machine has already passed through that moment. The Time Traveller offers the analogy of the difficulty of seeing a speeding bullet traveling through the air.

But how much of this discussion actually makes sense? (We certainly would argue that it makes for a great read!) As for the Time Traveller's argument that "Time is only a kind of Space," it is certainly true that our *perception* of time is very different from our perception of space. The notion of what it means to move through space, and even to move through space at different rates, makes some intuitive sense to us. Our "rate of travel through space," our speed, is the distance traveled divided by the time interval required to cover that distance (i.e., in the simple case of straight-line motion at constant speed). The units by which we measure "rate of movement through space" are units of distance divided by units of time. Thus 60 miles per hour is a faster rate of movement through space than 30 miles per hour.

How can we characterize the "speed" or "rate of movement" through time? Suppose we say something like 1 hour per second, so 1 hour per second would be 3,600 seconds per second. The problem is that we have the same units in both the numerator and the denominator of our quantity, so they cancel out and we end up with an answer of simply "3,600," a pure number. So what does this mean, 3,600 "what"?

In fact, our previous discussion really involves two different times. One we might call external time and designate it t. This is the time by which most of us, excluding the Time Traveller, live our lives. One can think of it as based on the time measured by an atomic clock located at the National Institute of Standards and Technology in Fort Collins, Colorado. Many other clocks are synchronized to this by radio signals. The second time that enters the discussion is the Time Traveller's own personal biological clock time, or pocket watch time, proportional, for example, to the number of heartbeats or the number of ticks of his watch that have occurred since some agreed-on starting point. Let us call this time T. In the usual situations t and T are at least roughly the same (although the rate at which a person's heart beats is somewhat variable). We can say that normally, $t/T = 1$ sec (of external time) / 1sec (of personal time).

When the machine, with the Time Traveller inside, travels into the future, t will be greater than T. That is, a long time must go by in the outside world while the Time Traveller ages only a little bit. For example, let's say that the Time Traveller spends one minute, according to his personal time, in the time machine ($T = 1$ minute). Then suppose that when he steps out of the machine and looks at the daily paper, he finds the date is one year later than when he started his trip. He has traveled one year (more precisely, one year minus one minute) into the future, and we can say that his "rate of travel," t/T, is equal to 1 year of external time / 1 minute of personal time. If we do not specify that these are two different times, then the notion of "rate of time travel" becomes rather confusing. This is because, as we discussed earlier, we could specify the numerator and denominator in the same units, for example, seconds, and then t/T would be just a pure number whose meaning is hard to interpret.

The notion that the machine would be invisible as it travels doesn't make sense. If the machine is traveling, into the future for example, then it will be continually present and thus constantly visible to the Time Traveller's guests. In order for the machine to age only a few minutes while years pass by outside the time machine, all processes within the time machine, including the physiological processes of any time traveler, must seem to happen very slowly. To external observers, the Time Traveller and his machine appear frozen in place.

Conversely, the Time Traveller will see things in the outside world happening at a highly accelerated rate, since he will see a year's worth of events crammed into a minute. Wells's fiction depicts this correctly. In the following passage, the Time Traveler describes the view from inside the machine during his trip into the future:

> The jerking sun became a streak of fire, a brilliant arch, in space, the moon a fainter fluctuating band . . . Presently I noted that the sun belt swayed up and down from solstice to solstice in a minute or less, and that consequently my pace was over a year a minute, and minute by minute the white snow flashed across the world and vanished, and was followed by the bright, brief green of spring.

Our earlier conclusion that the machine must be constantly visible to external observers implicitly assumes that the machine travels *continuously* through time. By this we mean that in order to go from moment A to moment B, the machine must pass through all the moments in between. Let us now consider the possibility that the machine time jumps discontinuously through time. This idea as applied to Wells's time machine is ruled out by the law of conservation of energy. The mass of the time machine and the energy it represents by virtue of the famous Einstein relation $E = mc^2$ cannot simply disappear, since the total energy in the universe is conserved, that is, remains constant, in time. (As a result of Einstein's relation, we will often use the terms "mass" and "energy" interchangeably.) Suppose that an external observer sees the Time Traveller get into his machine, turn it on, and disappear. As far as the external observer is concerned, the energy of the Traveller and his machine have disappeared from the universe, with no compensating increase in energy elsewhere in the universe to make up the difference. Likewise, an external observer who sees the time machine and its occupant appear out of nowhere will see an increase in the energy of the universe with no compensating decrease anywhere else.

There is, however, another version of this idea, which we will explore in detail later. It involves the Time Traveller taking an alternate path into the past or future through a "wormhole." While in the wormhole, the Time Traveller would be invisible to those outside and would reemerge at a different time. That's probably not what Wells was thinking of, since wormholes hadn't been imagined yet. When the Time Traveller enters the wormhole time machine, he disappears from the external universe, but *the mass of the wormhole increases* by an amount equal to the Time Traveller's mass. So an external observer will say that mass (energy) is conserved. Similarly, when the Time Traveller exits from the other end of the wormhole, external observers will see the mass of the wormhole decrease by an amount equal to the Time Traveller's mass. So for each set of external observers, the mass (energy) of (Time Traveller + wormhole) remains constant. We will explore in more detail some of the subtleties of energy conservation associated with this method of time travel in a later chapter.

Incidentally, the existence of conservation laws, which state that there are various properties of a system that remain constant in time, is one indication that there are important distinctions between time and space. This is in contrast to Wells's statement, quoted earlier, about the lack of such distinction. There are no corresponding laws concerning quantities remaining constant in space. It is true that relativity, as we will see, shows that space and time are much more interconnected than was previously thought, but the laws of physics also distinguish between them.

The Time Traveller implies that the machine occupies the same space but only travels through time. What exactly does it mean to say that an object "stays in the same location in space?" Well obviously, you say, the machine doesn't move around on the table. But the table and the Time Traveller's house are sitting on the surface of the earth. The earth is rotating on its axis and revolving around the sun, therefore so is the time machine. Since the earth does not "stay in the same location in space," what does it mean to say that the time machine does? If we assume, as Newton did, the existence of an absolute space against which all motion can be gauged, then from our previous argument it seems very unlikely that the earth could always be at rest relative to this "absolute space." (*Relative*—now that's a word we're going to hear a lot in our discussions.)

When we say something "stays in the same place" or is "at rest," we are implicitly assuming the additional phrase "with respect to, or relative to, something or other." For example, if an observer is riding in a car traveling at 60 miles per hour, the car and observer are traveling at this speed relative to the ground. However, the observer's speed *relative to the car* is zero! So he can equally truthfully say that he is moving or that he's staying in the same place.

It all depends on what the observer is using for his points of reference. If we say that the time machine remains at its same location in this absolute space, then the Time Traveller will be in for a surprise. He will find that the surface of the earth will move out from under the time machine, leaving it hanging in the vacuum of space. If that's the case, he'd better be careful about when he turns off the machine.

Let us suppose that the time machine does make a jump from one point in time to another. Already the specter of time travel paradox begins to emerge, as nicely described in an article by the philosopher Michael Dummett. Suppose that on Sunday at 12:00 noon, the Time Traveller places the miniature model time machine on the table and sends it off on its journey to the day before, Saturday, at 12:00 noon. Then anyone coming into the room on Saturday after 12:00 noon would have seen the time machine on the table. But then it would seem that when the Time Traveller comes into the room on Sunday, carrying the machine, he will see a "copy" of the machine already on the table. The copy on the table will be the machine that traveled back (i.e., the one he is about to send) to the past to Saturday and which has been sitting on the table ever since. But the copy is already occupying the place where he intends to put his machine.

To avoid the problem of the two machines getting in each other's way, let us suppose instead that when the Time Traveller first comes to the table on Sunday, he finds it empty. He places his model on the table and sends it off. Where, then, did it go (in space as well as in time) if it was not on the table when he came in? It appears that someone or something must have moved the machine in between the time that it appeared on the table on Saturday and the time that the Time Traveller placed his model there on Sunday. Perhaps the housekeeper placed it back in the Time Traveller's lab on Saturday at 1:00 p.m. to avoid having it damaged. On Sunday, the Time Traveller goes to his lab, picks up the model machine and takes it to the living room where he places it on the table.

There are several curious things about this latter scenario. Suppose the housekeeper decides not to move the machine but to leave it on the table. Then we would have a consistency problem (Dummett discusses one way around this). If we assume that she in fact *must* move the time machine, then the actions of the housekeeper on Saturday (i.e., whether she moves the machine or not) are determined by whether or not the Time Traveller chooses to turn on the machine on Sunday. So events in the past can be constrained by whether or not a time machine will be activated in the future. We could take this to extremes and say that an experiment I do today might be affected by the fact that someone is going to build a time machine a thousand years from now! This seems quite bizarre, because in science we are used to the idea that in performing an experiment, we are free to set things up (i.e., "choose our initial conditions") any way we like. Indeed, our whole process of science is in some ways predicated on this idea.

A second problem with our scenario is the following. Suppose that the Time Traveller places a tiny celebratory bottle of champagne on the seat of the model time machine, which he uncorks just before turning the machine on. The Time Traveller sets the machine off on Sunday, whereupon it effectively appears instantaneously on Saturday. Then if the housekeeper places the machine in the Time Traveller's lab, which sits there until he picks it up and takes it to the living room table on Sunday, he notices that there is a flat bottle of champagne on the seat of the machine. So the time machine that he places on the table cannot be exactly the same as the one he sent back. The one he sent back had a fresh bottle of champagne on the seat but the one he finds in his lab and subsequently places on the table has a bottle of flat champagne. If you say, "Well, the Time Traveller simply removes the stale bottle and replaces it with a fresh one before activating the machine," then you have the problem of explaining where the stale bottle came from in the first place. We will have more to say about this kind of paradox and its relation to something called the "second law of thermodynamics" later in the book.

Time and Space Measurements

After our brief foray into time travel (which was meant to whet your appetite), let us consider the more mundane question of how we measure the position of an object in space and time. For our purposes, we will take a very practical approach and consider time to be "that which is measured by a clock." A clock is just a device that keeps going through repetitive cycles, for example, the swinging of a pendulum or the vibration of a mass on the end of a stretched spring. One then defines the length of a time interval as being proportional to the number of cycles.

Good clocks should be easy to reproduce exactly, and their rate of vibration should not be affected by external conditions. The periods of two pendulums will be different unless they have exactly the same length. And even if they do, the period will change slightly—but measurably—if the temperature changes, because that would cause the length to change slightly. Human hearts are obviously very bad clocks, since they beat at different rates for different people, and people are notoriously hard to reproduce exactly. And even for a particular person, the heart rate is different at different times, depending on whether they are asleep or running a marathon. The most accurate clocks today are atomic clocks, which are based on the vibrations of light waves emitted by atoms, often atoms of the element cesium. These make good clocks because any two of the cesium atoms used are absolutely identical, and their rate of vibration is affected only by very extreme changes in external conditions. Such clocks can measure time to accuracies of billionths of a second or better.

By contrast, we can determine positions of objects in space using a series of objects of fixed length, such as meter sticks. Suppose that lightning strikes the roof of a train station at 1:00 p.m. We will call the lightning strike "an event." To locate the event in space and time, we need four numbers, or four "coordinates"—the spatial coordinates in terms of a set of X, Y, and Z spatial axes—and the time at which the event occurred. But first we need to choose a set of fixed axes to measure the spatial coordinates. We can choose these axes to be fixed with respect to the ground or with respect to a speeding train, car, or rocket. Once we have chosen our axes, we can imagine laying out a grid or "jungle gym" of meter sticks along each of the three axes and at rest with respect to them and to each other. The spatial location of an event is denoted by the x, y, and z coordinates along the three axes, as measured using the grid of meter sticks. To measure the time at which an event occurs, we imagine a "latticework" of points in space with a clock placed at each point in the lattice. The time at which we deem an event to occur will be the time reading on the clock nearest the event. For this setup to make sense we must *synchronize all the clocks* with one another. It turns out that there are subtleties associated with this process, which we will analyze carefully in the next chapter. This network of meter sticks and synchronized clocks is called a "frame of reference." The fact that the spatial and temporal positions of an event are measured in different ways is another signal of a physical distinction between space and time. The procedure by which quantities are measured is important because physics is ultimately an experimental science.

There are certain kinds of reference frames that can be singled out for discussion. We have all had the experience of falling asleep on a train while waiting for it to pull out of the station and then suddenly waking up and looking out the window at a train on the other track. If the motion is smooth, with no bumps, and no changes of direction (i.e., "constant velocity," or motion in a straight line at constant speed), then we cannot tell whether it is our train or the other that is moving. If we drop or roll balls on the floor of the train car, they will behave in the same way, whether it is our train or the other that is moving. A frame of reference that is attached to such a train in which we cannot distinguish rest from uniform motion is called an "inertial frame."

The name "inertial frame of reference" comes from Newton's first law of motion. This law says that "an object at rest remains at rest, and an object in motion continues in motion in a straight line at constant speed, unless acted on by an external force." In plainer but somewhat less precise language, Newton's first law says that if left alone an object will tend to continue doing whatever it's doing. Frames of reference in which objects behave this way are called inertial frames; frames in which they don't are called noninertial frames.

An air table is a device used in elementary physics labs. It consists of a horizontal table with many tiny holes drilled in the surface through which a constant stream of air is blown. A light hockey puck placed on the table will move essentially without friction. If placed at rest it will remain at the same spot on the table. If given a shove, it will move at constant speed in a straight line until it hits the edge of the table. Now consider two additional identical air tables. Place one in a car moving at constant speed in a straight line relative to the lab containing the original air table. Place the second air table in a car that is accelerating (i.e., whose velocity relative to the lab frame of reference is increasing). Let us assume that the windows of the car have been blacked out so that the passengers cannot see outside. (Don't try this at home!) Hence, they can only make conclusions regarding their motion from observations made from within the car. The frame of reference attached to the first car is an inertial frame of reference. This is because a hockey puck placed on the air table in that frame will continue doing whatever it was doing. If initially at rest it will remain so; if moving it will continue moving in a straight line with constant speed. In other words, it behaves according to Newton's first law, just like the hockey puck on the air table back in the lab. However, consider the placement of a hockey puck on the air table in the accelerating car. If placed at rest on the table, it will not remain at rest, but will slide backward (if the car is accelerating forward in a straight line). To an observer in the car this seems peculiar, because there is no obvious external force acting on the puck, since the air table is frictionless. Yet the puck does not obey Newton's first law. An observer in this car will notice that they too feel pushed back in the seat by some unseen force. Similarly, a hockey puck placed on an air table on a rotating merry-go-round will feel a peculiar force that makes it move in a curved, rather than a straight, path if launched from the center outward to the edge along a radius. So we can tell the difference between inertial and noninertial motion. More generally, an inertial frame is one which is nonaccelerated and nonrotating (actually, rotation is an example of accelerated motion), as seen from another inertial frame.

How do we relate the measurement of an event in one frame to measurements of the same event in another frame? For two inertial frames, there is a simple intuitive relationship between the coordinates of an event in one frame and its coordinates in the other frame. This set of relations is called the "Galilean transformations," named after the famous seventeenth-century Italian physicist Galileo Galilei, who laid the framework for the study of motion.

Suppose that we have two frames of reference that move at a constant speed along a straight line relative to one another. For example, suppose one frame is at rest with respect to some train tracks and the other frame is at rest with respect to (or "attached to") a train moving at constant speed along a straight stretch of the track. For simplicity, let us consider the relative motion to be only along the x axis. A firecracker goes off on the tracks at position x,y,z at time t, as measured in the track frame. The train moves along the positive x axis with constant speed v. What are the coordinates of the same event in the train frame?

Since the relative motion is only along the x axis, the y and z coordinates should be the same in both frames. We will also make the (obvious, you say?) assumption that time is the same in both frames, so that the time coordinates of the events are the same. All that remains is to determine the relation between the x coordinates of the events in the two frames. (Incidentally, we could assume an arbitrary direction of relative motion, but that would just complicate the equations without adding much to our understanding in the present discussion.) Let us arbitrarily call the x coordinate of the event relative to the track frame, which we will call the "S(track) frame," simply x. The corresponding coordinate of the same event in the train frame, which we call the "S'(train) frame," will be denoted as x'. In figure 2.1, the S(track) and S'(train) frames are shown; the origins of the coordinate systems in each frame are denoted by O and O', respectively, and coincide with one another, that is, they are just passing one another, when $t = 0$. The coordinate axes of the S(track) frame are designated X and Z; those in the S'(train) frame are denoted X' and Z' (for simplicity, we have suppressed the Y, Y' axes in figure 2.1). The S'(train) frame moves with constant velocity v to the right along the X and X' axes relative to the S(track) frame. (Note that a velocity has both a speed, i.e., a size or magnitude, and a direction.) At time t in the S(track) frame the firecracker explodes [in this discussion we assume that the time of explosion is the same in the S'(train) frame, namely, $t' = t$] at location x', y', z'. Since the relative motion is only along the X and X' axes, the y and z coordinates are the same in both frames, that is, $y' = y$ and $z' = z$. We see from the diagram that the corresponding x coordinate of the explosion is simply $x = x' + vt$, namely, its position in the S'(train) frame plus the

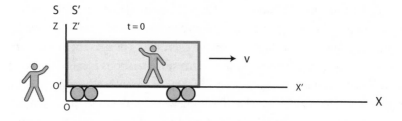

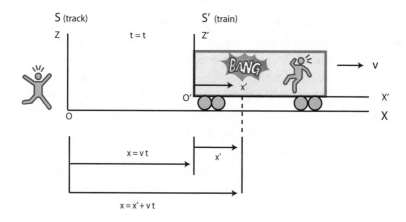

FIG. 2.1. Observers in two inertial frames. The frame S(track) is attached to the train track, and the frame S'(train) is attached to a train moving at constant velocity.

horizontal distance which the origin of the S'(train) frame has moved during the time t.

Therefore, the set of relations between the coordinates in the S'(train) and S(track) frames can be written (after a minor rearrangement) as:

The Galilean Coordinate Transformations

$$x' = x - vt$$
$$y' = y$$
$$z' = z$$
$$t' = t$$

These are called the Galilean transformations. Let us again emphasize the important point that x and x', for example, represent the coordinates of the *same* event (the explosion of the firecracker in this example) as seen from *two different reference frames*. They do not refer to two different events. It will be important to keep this in mind during much of the subsequent discussion. The velocity v

can, of course, be directed to the left, that is, in the negative *x* direction. In that case, *v* would be replaced by −*v* in the transformation equations, and the arrow labeled *v* in figure 2.1 would point to the left.

In our previous example, the firecracker was at rest in the S'(train) frame prior to the explosion. Now consider another example in which an object is moving relative to both frames. Referring to our previous figure, let the object move with speed *u′* to the right, as measured in the S'(train) frame. The same object is measured to have speed *u* to the right, as measured in the S(track) frame. How are these two velocities related to one another? If you guessed that there is also a Galilean transformation for velocities, you'd be right. [Note that *v* still represents the velocity of *reference frame* S'(train) relative to S(track) as before. We have now introduced a second velocity *u* which represents the velocity of the as yet unspecified *object* relative to S(track), and *u′* the object's velocity relative to S'(train).]

To make things concrete, let's suppose that the object is a person who walks at a speed of *u′* = 1 mph to the right with respect to the floor of the train, that is, as measured in frame S'(train). (Once again, for simplicity, we will consider all the motion to be along the *x* and *x′* axes.) Let the speed of the train with respect to the track, that is, the speed of the S'(train) frame relative to the S(track) frame, be *v* = 60 mph. How fast is the person on the train moving *relative to the track*? It's fairly easy to see that the speed of the person relative to the track (*u*) will be equal to the speed of the person with respect to the train (*u′*) + the speed of the train with respect to the track (*v*), namely, *u* = 1 mph + 60 mph = 61 mph. More generally, we have *u* = *u′* + *v*.

Another simple example is the case, experienced by many people nowadays, of walking along a moving walkway. If the walkway moves, for example, at a speed of 2 feet per second relative to the ground, and you walk at a speed of 3 feet per second with respect to the walkway, then your speed relative to the ground is 5 feet per second.

If in our expression, *u* = *u′* + *v*, we instead write the primed quantities in terms of the unprimed quantities, as before, we have:

The Galilean Velocity Transformation

$$u' = u - v$$

The velocity transformation can be easily gotten from the Galilean coordinate transformations. The reader who is interested in these details can find them in appendix 1.

The Galilean transformations are simple and intuitively obvious. As we will see in the next chapter, they are also wrong.

3

Lorentz Transformations
and Special Relativity

Nothing puzzles me more than time
and space, and yet nothing puzzles me less,
for I never think about them.

CHARLES LAMB

It gets late early out there.

YOGI BERRA

In this chapter we will look at how experiments force us to modify the simple—and
seemingly obvious—Galilean transformations (introduced at the end of the
chapter 2) when we deal with objects and reference frames whose speeds are
comparable to c, the speed of light. These modifications will lead us to Einstein's special theory of relativity. Since light and the speed of light will be so
important in this story, we'll begin with a brief look at the state of knowledge
which physicists had about this subject in the years leading up to Einstein's
accomplishment.

For nearly two centuries after the time of Newton, physicists debated
whether a beam of light was a stream of particles or whether it was a wave,
similar to ripples on the surface of a pond. In the case of a wave, one has some
medium, for example, the water in the pond, which oscillates or vibrates as the
wave passes. In the case of the water wave, the water molecules oscillate up and
down as the wave moves, let us say, from left to right. At a given moment, the
water molecules at a particular point in the pond, call it point P, may be at their
maximum height. If we were watching the wave, we would say that at that mo-

< 22 >

ment there was a crest of the wave at point P. A bit to the right of P, the water molecules would be momentarily at the lowest point of their oscillation, and there would be what is called a wave trough. A little later, the water molecules at P will be at the lowest point of their cycle, so there will be a trough at P, while the crest which was there initially will have moved to the right. Note that it is the wave itself, that is, the shape of the surface that moves from left to right. The water molecules themselves do not move from left to right with the wave, but just bounce up and down in place. A similar situation occurs in the case of sound, but in that case molecules in the air oscillate back and forth as a sound wave passes, rather than up and down.

When waves come from two different sources (e.g., spreading out from two different openings in a breakwater into the otherwise smooth surface of the harbor behind), the waves can exhibit a phenomenon called "interference." This occurs, for example, when crests from the two waves arrive at the same point at the same time, giving rise to crests that are twice as high as those from the individual waves. That is, at those points the water molecules reach twice the height during their up-and-down oscillation than they would if only one of the waves was present. Similarly, if troughs from the two waves arrive together, they produce a trough that is twice as deep as those of either wave by itself. At such points the two waves are said to interfere "constructively." On the other hand, there will be points where crests from one wave, tugging the water molecules upward, arrive at the same time as troughs from the other wave, tugging downward. The result is that the water molecules never feel any net force, up or down. Thus, at those points the water doesn't oscillate at all, and the surface remains still. At these points the waves are said to interfere "destructively." In between these two kinds of points one sees, as you would expect, water oscillations that are not totally absent but are not as vigorous as at the points of complete constructive interference. Interference is a phenomenon characteristic of waves, and its occurrence is a sure indication of the presence of wavelike behavior.

In 1801, the English physicist Thomas Young passed a beam of light through two parallel slits in a screen and observed an interference pattern of alternate bright and dark bands on a second screen behind the slits. Such patterns are much harder to see in the case of light waves than that of water waves because the wavelength (the distance between successive crests or successive troughs) of a light wave is something like a million times shorter than that of water waves. This turns out to mean that very narrow slits must be used in the case of light. Young's experiment indicated conclusively that light had a wavelike nature. (About a century later, with the advent of quantum mechanics, it was discovered that light also has particle-like properties, but this need not concern us at the moment.)

James Clerk Maxwell's Great Idea

While Young's experiment seemed to settle the question that light was a wave, it left other questions open. What, exactly, was it that was oscillating as a light wave passed, and in what medium was it propagating? The first of these questions was answered in the second half of the nineteenth century by the work of the Scottish physicist James Clerk Maxwell on the theory of electromagnetism, that is, the combined theory of electricity and magnetism, which turned out to be intimately related to one another. Through the work of physicists such as Coulomb, Ampère, and Faraday, a set of equations governing what are called electric and magnetic fields were derived. These fields describe the electric and magnetic forces that act on electrically charged particles in various situations. Maxwell noticed that the equations for the electric and magnetic fields were rather similar, but that there was a term in one of the equations for the electric field which had no counterpart in the corresponding equation for the magnetic field. Although at that time there was no experimental evidence for this latter term, Maxwell guessed that it should be there.

When Maxwell included this new term he found that the enlarged set of equations had a remarkable new kind of solution. This solution corresponds to waves composed of oscillating electric and magnetic fields, propagating through space similarly to water waves through water. Moreover, he calculated the velocity of these waves in terms of two parameters that described the strength of the electric and magnetic forces between given configurations of electric charges and currents. The value of these parameters was known from measurements of these forces. When Maxwell plugged in the known values of these parameters, he found that the speed of these new waves, which are now called electromagnetic waves, was predicted by the equations to be 300,000 kilometers per second, that is to say, about 186,000 miles per second—the speed of light waves!

It was inconceivable that this could be a coincidence, and the obvious conclusion was that light waves were, in fact, examples of this new kind of wave that the equations of electromagnetism, with Maxwell's term added, predicted. The exclamation point at the end of the preceding paragraph is well deserved. This is one of the most remarkable and beautiful results in the history of theoretical physics. Maxwell was able to predict the speed of light, the quantity we now call c, in terms of two well-known constants that, before his theory, appeared to have nothing at all to do with light waves. One might guess that, when he first calculated the speed of the new kind of waves predicted by his equations and saw the answer, he felt an exhilaration comparable to that felt by a major league ball player who has just hit a walk off grand slam home run in the seventh game of the World Series. Because of this remarkable result that followed from Maxwell's contribution, the entire set of four equations governing the electric and magnetic fields are now called Maxwell's equations, even though he was only personally responsible for the form of one of them.

As we have emphasized, when you talk about the velocity of an object, you must always be clear—velocity with respect to what? If we say the speed of sound is about 300 meters per second, we mean, although we do not always say, that this is the speed of sound relative to the air, one of the media through which sound waves propagate. So what about light? When Maxwell predicted that the speed of light was c, that is, about 3×10^8 meters per second: to what was this relative? Since waves need a medium in which to propagate, and no such medium was apparent in the case of light, one was invented, and given the name "aether" (pronounced "ether"). The aether was pictured as a kind of massless, colorless, and otherwise undetectable fluid whose one mission in life was to provide a medium in which light waves, that is, Maxwell's electromagnetic waves, could propagate. (Obviously the word "aether" in this usage has nothing to do with the drug which can be used to induce anesthesia.) So, by analogy with sound, c was presumed to be the speed of light relative to the aether. Or, to put it another way, it was the speed of light in a very special (or as physicists say, "preferred") reference frame, namely, the reference frame that was at rest relative to the aether. Unfortunately, since no one could see, feel, hear, taste, nor smell the aether, that presumption was a little hard to verify.

The Michelson-Morley Experiment

But one could do something that was almost as good, or so it appeared. Two American scientists, Albert Michelson of Case Institute of Applied Technology and Edward Morley of Western Reserve University (the two neighboring suburban Cleveland institutions have since combined to form today's Case Western Reserve University) set out to do it in 1887. To a physicist, the earth plays no particularly special role. Therefore, Michelson and Morley had no reason to believe that the frame of reference in which the earth was at rest at any particular moment was the preferred frame defined by Maxwell's equations, that is, the frame of reference of the aether. Thus, they expected that the speed of the earth's reference frame relative to the aether would be at least as great as the speed of the earth in its orbital motion around the sun.

We should note that the earth itself does not, strictly, constitute an inertial frame, because it is not moving with constant velocity. A reference frame attached to the center of the earth is accelerating, because the direction of its velocity is continuously changing as it follows its (nearly) circular path around the sun. In addition, a point on the surface of the earth has an additional acceleration due to the earth's rotation on its axis. These accelerations are both relatively small, compared, for example, to the acceleration of Newton's famous falling apple, and it is often a reasonable approximation to regard the earth itself as defining an inertial frame of reference. An excellent approximation to an inertial frame is a frame attached to the center of the sun and with its axes pointing in a fixed direction relative to the distant stars, so that the axes are not rotating.

Let's call the earth's orbital speed v. The earth's orbit is roughly a circle whose radius, r, is about 93,000,000 miles, or about 1.5×10^8 kilometers. In one year, which turns out to be about 3×10^7 seconds, the earth travels a distance equal to the circumference of the orbit, $2\pi r$. This yields a value of about 30 kilometers per second for v. In everyday terms this is a very high speed, about one hundred times the speed of sound, but it is only a very small fraction (about one thousandth) of the speed of light.

Michelson and Morley set out to demonstrate the existence of a preferred frame for light waves by measuring the earth's velocity with respect to it. Suppose that at some instant of time the earth, in its circular motion, is moving almost directly away from some particular star. Given the number of visible stars, that's pretty much guaranteed to be the case. Consider the speed of light as seen from the earth. To do this, we'll go back to our discussion of the Galilean velocity transformation equations in the preceding chapter. Only this time, instead of letting the two reference frames, S and S', represent the train and track frames for a moving train, we'll let S(aether) represent the reference frame of the aether, and S'(earth) the reference frame in which the earth is momentarily at rest.

To continue, in our S(aether) and S'(earth) frames, we take the earth to be moving along the x (and x') axis relative to the aether. Then the v in the Galilean velocity transformation equations will be the speed of the earth relative to the aether, and u will be the speed of the light relative to the aether in the direction in which the earth is moving, that is, in the x direction, so that $u = c$, (that's what defines the aether) and u' in the transformation equations will be the speed of the starlight relative to the earth in the x direction. The Galilean velocity transformation equation $u' = u - v$ becomes $u' = c - v$. That is, the speed of the starlight relative to the earth along the x axis, which is also the line from the earth to the star, is predicted to be a little bit less than c, because the earth is "running away" from the starlight with speed v. We can recast this equation as $v = c - u'$, where v is the velocity of the earth through the aether, which Michelson and Morley wished to measure. Remember their guess was that v might be about equal to the earth's orbital speed of about $0.001c$. Since they guessed that v was probably going to be much less than c, they performed very careful measurements in order to get a believable value for v.

Before we can understand the experiment, we must also remind ourselves of one other aspect of the Galilean transformations. Let's look at the difference between the two reference frames in the rate at which an object, or in our case, a light pulse is moving along the y or z axis in the aether and in the earth frames, that is, in a direction perpendicular to the velocity of the earth. Here, the Galilean transformations, as well as our common sense, tell us the difference is zero. However, the speed of a light pulse moving along the y or z axis will be affected by the earth's motion in the x direction, since the speed in a given frame depends on the rate of motion of the pulse in both the x and y directions in that frame. This is analogous to the fact that a boatman rowing cross-stream against a moving river must have part of his motion through the water directed against the current in order to end up on exactly the opposite side of the riverbank. Part of his motion must fight the current. Hence, his speed in the direction perpendicular to the bank will not be as great as if there were no current.

Michelson and Morley admitted a beam of light into their apparatus, called an interferometer. This is illustrated in figure 3.1. The light was initially traveling perpendicular the earth's direction of orbital motion.

Then, by use of a beam splitter, oriented at 45° to the light path, they broke the light into two beams. One was transmitted through the beam splitter, and continued a distance d perpendicular to the direction of the earth's motion in *the earth (primed) frame*, that is, along the y', not the y, axis in order to hit the first mirror. Since the y' axis itself is moving along with the earth at speed v, this beam has a velocity v in the x direction in the aether frame. Since, by definition,

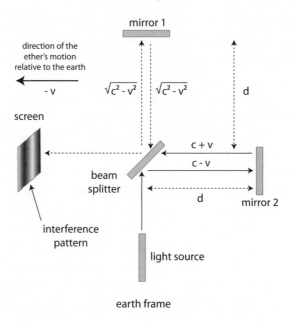

FIG. 3.1. The Michelson-Morley experiment. A beam of
light is split into two parts. One beam moves at right angles
to the direction of the earth's motion through the ether; the
other moves first against and then with the earth's motion.
The two beams then recombine at the screen on the left.

light moves with speed c in the aether frame, the Pythagorean theorem tells
us this beam will have a speed we'll call $v_y = \sqrt{c^2 - v^2}$ along the y axis.[1] But
since by either the Galilean transformations or common sense the velocities in
directions perpendicular to the earth's motion are the same in either S(aether)
or S'(earth), the light pulse will move with speed $v_y' = v_y$ along the y' axis. There
it was reflected by mirror 1 back to the beam splitter.

The other beam was reflected off the beam splitter but traveled the same
distance d sidewise and was then likewise reflected from mirror 2 back to the
beam splitter. A portion of the two beams then recombined and went to the
left, where they both hit a screen and formed an interference pattern. Both had
traveled a distance $2d$. If the two beams traveled at the same speed they would
take an equal amount of time to make their respective journeys, and they would

---

1. To use some sailing terminology, you can think of this as a result of having to "tack" cross-
wise against an aether "current".

show perfect constructive interference. "Crests" (points where the fields had their maximum values) of the two beams would arrive back at the same time and reinforce one another, as would "troughs," so that the two beams interfere constructively.[2]

But this was not what Michelson and Morley expected to see, because they did not believe the two beams traveled at the same speed. A bit of algebra (which we won't go into) shows that the "up-and-down" beam in figure 3.1 always beats the "side to side" beam. The difference in travel time for the two beams is observable, since the interference is no longer exactly constructive. The size of this effect should have allowed Michelson and Morley to obtain a value for the earth's speed, v, through the aether.

What happened when they did the experiment? Michelson and Morley found that, within the accuracy of their measurement, $v = 0$. Taken at face value, this meant that at the time of their measurement the earth happened to be at rest relative to the aether, an almost inconceivable coincidence. But anyway, it was easy to check that idea. They just had to redo the experiment six months later when the earth, as it went around its circular orbit, would be heading in exactly the opposite direction. If the earth happened to be at rest in the aether's frame of reference at the time of the first measurement, six months later its velocity relative to the aether would be different. However, when they repeated the experiment, Michelson and Morley got the same result. The light beams moving parallel to and perpendicular to the direction of the earth's orbital velocity appeared to have the same velocity relative to the earth. Now the result could not be attributed to any coincidence, however improbable. Assuming Michelson and Morley had done their work correctly, there was no escaping a conclusion that was difficult to accept. The commonsense procedure for adding velocities, embodied in the Galilean transformations, doesn't work in the case of light! If a light beam moves through space at speed c and an observer moves through space at speed v, the observer also sees the light beam moving by him at speed c.

The Michelson-Morley experiment is one of the truly seminal experiments in the history of physics. Like all important experiments, it has been redone many times by others to verify the result. The experiment is a difficult one to do

2. An important point in the design of the apparatus was that when they were detected, both the forward-and-back and left-and-right beams had passed once through the half silvered mirror, and had been reflected once off it. Hence, the different speed of light in glass than in air canceled out between the two beams and produced no difference in travel time between the two beams.

because of the small size of the effect expected, which is just about at the limit of what the experiment is capable of detecting.

The Two Principles of Relativity

Einstein's special theory of relativity, published in 1905, rests on two basic principles from which everything else follows. The first principle of relativity states that all physical laws have the same form in every inertial frame. Since inertial frames differ from one another by being in motion with constant velocity relative to each other, the first principle says that if you are in a closed room, there is no physics experiment that you can do that will tell you whether you are at rest or in motion with a constant velocity. In fact, it says the question of whether you are at rest or in uniform motion isn't really meaningful, because the laws of physics do not pick out any particular inertial frame as distinguishable from all the others; physicists would say there is no "preferred" inertial frame. Thus there is no unique way to answer the question, "in uniform motion relative to what?" You're always entitled to regard your own inertial frame as, so to speak, the "master frame" relative to which velocities are measured.

Maxwell's equations leave open the possibility that light travels with speed c relative to the source of the light, for example, the bulb of some lamp. The second principle of relativity, as adopted by Einstein, is that the speed of light doesn't depend on the motion of the body emitting the light. There were experiments known at the time (which we will not go into here) in support of this principle. If the speed of light doesn't depend on the motion of the emitting body, there is nothing else on which it can depend without violating the first principle. The two principles of relativity together imply that observers in all inertial frames measure the speed of light to be c relative to their reference frame.

The Michelson-Morley experiment provides evidence that, as an experimental fact, the speed of light is the same in all inertial frames. While Einstein was aware of the Michelson-Morley experiment, he seems, perhaps, to have based his own thinking more on a strong intuitive conviction that Maxwell's equations for electromagnetism should be valid in every inertial frame, and not just valid in some preferred frame picked out by an otherwise unobservable aether.

The validity of the first principle of relativity, like all physical principles or laws, rests on experiment. However, it places very strong constraints on the possible forms that physical laws can take, and so far we've never observed those constraints being violated. One powerful example of the result of such constraints occurs in the case of one of the most important of all physical laws—that of conservation of energy. It turns out that, in the form it was known before special relativity, it did not obey the first principle of relativity and looked different in different inertial frames. Einstein suggested that a proper formulation of the law of conservation of energy ought to be constrained by the first principle of relativity. The proposed revisions led to a number of experimental predictions, including the famous equation $E = mc^2$. These predictions have been tested extensively in many different experiments and so far have passed all the tests. In fact, these predictions as to the form of various physical laws provide much stronger experimental support for special relativity than does the prediction of the universal value of the speed of light in all inertial frames. That prediction rests on the Michelson-Morley experiment and various successors, which are difficult to perform with a high level of precision.

The Lorentz Transformations

It follows from the outcome of the Michelson-Morley experiment and from Einstein's first principle of relativity that the Galilean transformations cannot be completely correct and must be modified in situations where the speeds u or v become close to c. The modification, however, must be such that the Galilean transformations remain valid in situations where the speeds involved are much less than c, where our everyday observations tell us they are correct. The first principle then says that, provided we transform our coordinates correctly in going from one inertial frame to another, all physical laws have the same form in every inertial frame.

One can actually find an alternative set of transformations that satisfy these requirements, and, in particular, give $u' = c$ when $u = c$. These are called the Lorentz transformation equations (Lorentz had developed these equations prior to Einstein, but he did not correctly grasp their physical implications). In appendix 2 we will discuss in more detail how these equations may be arrived at. Here we will simply write them down and examine their properties and consequences. We again suppose that we have two inertial reference frames, and take one to be the frame S(earth), in which the earth is momentarily at rest. Since we now wish to put aside the rather unphysical idea of the undetectable aether, we will take the other frame to be S'(ship), with its origin attached to a passing starship moving by the earth with constant velocity v. As before, we orient the two reference frames so that their axes are parallel, with v directed along the common x and x' axes. We will also set the clocks at the origins of the two frames so that observers on both the earth and in the ship see both clocks reading $t = t' = 0$ at the moment the two origins pass one another.

We remind the reader of the situation under consideration. Suppose we have an "event"—something that happens at a definite time and place, for example, a bat striking a ball. We can label the coordinates of this event by giving its time and space coordinates (t,x,y,z), as read on clocks and meter sticks at rest in S(earth). We can also label the position and time of the *same* event by giving its coordinates $(t',x',y',z',)$ in the frame S'(ship). The transformation equations then give the primed (ship frame) coordinates of an event in terms of its unprimed (earth frame) coordinates. We first recall the form of the Galilean transformations from chapter 2. The Lorentz transformations follow:

Galilean Transformation Equations
$$t' = t, \, x' = x - vt, \, y' = y, \, z' = z$$

Lorentz Transformation Equations

$$t' = \frac{t - \dfrac{vx}{c^2}}{\sqrt{1 - \dfrac{v^2}{c^2}}}, \, x' = \frac{x - vt}{\sqrt{1 - \dfrac{v^2}{c^2}}}, \, y' = y, \, z' = z$$

First, how do these equations behave when we are concerned only with speeds much less than the speed of light? For such cases, all the terms in the equations above which have a v in the numerator and c in the denominator will be very small, compared with the others, so we can safely ignore them (this should be especially true of the terms involving $\frac{v^2}{c^2}$, since when you square the already-small number $\frac{v}{c}$ you get a really, really small number). Now notice if we just throw away all the terms in the Lorentz transformation equations that involve $\frac{v}{c}$, you do indeed get back the Galilean transformations. The differences introduced by going to the Lorentz transformations become significant only when $\frac{v}{c}$ is not negligibly small.

In particular, the preceding remark applies to one of the most striking things about the Lorentz transformations. When we introduced the Galilean transformations, we just threw in the last equation, $t' = t$, as an afterthought, since there was no obvious reason why the time shown on a clock should be different just because the clock was moving. But that's no longer true if the clock is attached to a reference frame that is moving at a speed comparable to c. In that case, if you want to have the speed of light equal to c in both reference frames, it turns out that t' and t are necessarily different, and, in particular, that t' depends on both t and x. In other words, the time at which observers on the ship see an event occur depends not only on when it occurred in the earth frame, but also where. We shall see shortly just how this relates to the fact that the speed of light is c in both frames.

We have chosen the origin of the S′(ship) frame so that it is just passing the origin of the S(earth) frame when the clocks at the origin of both frames read zero. Also, the origin of S′(ship), where $x' = 0$, is moving with speed v relative to the earth. Hence, the point with $x' = 0$ should be at $x' = vt$. Looking at the first of the Lorentz transformation equations, we see it is indeed true that when $x' = 0$, $x = vt$, a property that is required if they are to make sense.

Finally, what about the speed of light in the two reference frames? Showing that the Lorentz transformations guarantee it is the same as seen by observers on earth and on the spaceship is just a matter of algebra. Let's suppose that at $t = 0$, we emit light pulses from the origin of the S(earth) reference frame in both the positive and negative directions along the x axis. Since light travels at speed c relative to the earth, the trajectories of the two pulses will be described by the equations $x = ct$ and $x = -ct$, respectively. We can summarize these two equations, after squaring both sides of each, by saying that the motion of the two pulses as seen by observers on earth satisfies the condition $x^2 - (ct)^2 = 0$. To show that observers on the spaceship also see the light pulses traveling at speed c we must ask whether the Lorentz transformations imply that it is also true that $x'^2 - (ct')^2 = 0$. In fact it turns out they imply a little more, namely"

$$x^2 - (ct)^2 = x'^2 - (ct')^2$$

for any value of $x^2 - (ct)^2$. *Almost everything we do in the rest of the book follows from this equation.* If you wish, you can just take our word that it is correct. You can also prove it yourself by substituting the Lorentz transformation equations for x' and t' into the right-hand side of the equation above. We give a proof of it in appendix 3.

The Invariant Interval

We're going to put what we've just told you (and what you've seen for yourself if you've been conscientious and done the algebra) in different language, which is convenient and commonly used. This also leads to an interesting partial analogy between the three-dimensional space of Euclidian geometry and the four-dimensional spacetime of relativity, that is, the set of all possible events. Let us define a quantity s^2, which we'll call the "interval" between an event at the origin of the reference frame S(earth), and an event with time and space coordinates (t,x,y,z) by the equation $s^2 = x^2 + y^2 + z^2 - (ct)^2 = r^2 - (ct)^2$. Here we've put the y and z coordinates back, even though they're pretty much just along for the ride, and made use of the three-dimensional generalization of the Pythagorean theorem, $x^2 + y^2 + z^2 = r^2$, to rewrite the equation in terms of the spatial distance r of the event from the origin.

What we've learned from our investigation of the Lorentz transformation is that $r^2 - (ct)^2 = r'^2 - (ct')^2$. That is, the interval has just the same form when expressed in terms of the coordinates in the ship frame. It is this property which gives s^2 its name of *invariant* interval. An invariant quantity is one that is the same in all inertial frames of reference, such as the speed of light, c. We refer to the transformation from S(earth) to S'(ship) by using the Lorentz transformation equations to relate the coordinates in the two frames, and say that s^2 is invariant under such transformations.

Let's consider for a moment purely spatial geometry. Instead of talking about transformations to a moving reference frame, we can discuss transforming to a new spatial coordinate system obtained by rotating the coordinate axes while keeping the axes mutually perpendicular. For example, in two dimensions, we might take new axes that connected opposite corners of the paper, instead of being horizontal and vertical. We'll call our new spatial axes in two dimensions, X' and Y'. [This is a new set of primed axes which have nothing to do with the reference frame S'(ship) and were obtained by a rotation, not a Lorentz transformation.] This situation is illustrated in figure 3.2.

We can also specify the position of a point P in the plane by giving its coordinates in the primed coordinate system. The primed and unprimed coordinates of the points will be different, but the combinations $x^2 + y^2$ and $x'^2 + y'^2$ will be equal, since our friend Pythagoras assures us that both are equal to r^2, where r is the distance of P from the origin. This distance certainly hasn't changed just because we chose to use a rotated set of coordinate axes. Thus, we say that r is invariant under rotations, because it has the same form in terms of the coordinates in two coordinate systems obtained from one another by a rotation. In simpler terms, we could say that the length of line r has an "existence" in the plane which is independent of any coordinate system we use. After all, we could have drawn the line on the page first, and then added the coordinate systems later.

We can think of s as being a kind of distance of an event from the origin in spacetime in the same way that r is the distance of a point in space from the

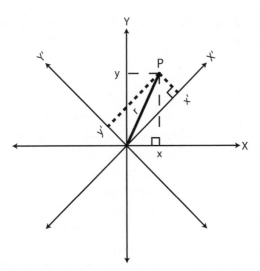

FIG. 3.2. A rotation of coordinate systems. Although the x and y components differ from the x' and y' components, the length of the line, r, is the same in both coordinate systems.

spatial origin. This analogy can be helpful, but it can't be pushed too far. In ordinary space, distances are always positive, but spacetime "distances" can be positive, negative, or zero. In the three-dimensional version of figure 3.2, $r^2 = x^2 + y^2 + z^2$, which is always positive (or trivially zero). Remember that the analog in four-dimensional spacetime, the interval $s^2 = r^2 - (ct)^2$, contains another piece in addition to r^2, and the term involving t has a different sign than the spatial terms. The minus sign is important, and is another example of the fact that while time and space are more closely related in special relativity than in Newtonian physics, as mentioned in chapter 2, they are not physically equivalent. In particular, in contrast with r^2, the invariant interval s^2 can be positive, negative or zero! For example, in the case of an event that occurs at the spatial origin, $r = 0$, and therefore whose only nonzero coordinate is t, $s^2 = -(ct)^2$, s^2 is negative. In the case of an event connected to the origin by a light signal, $r^2 - (ct)^2 = 0$, $s^2 = 0$.

In appendix 4 we will explore an approach that involves looking at Lorentz transformations to a different inertial frame as a kind of rotation of the coordinate axes in the x,t plane, rather than in a plane containing two spatial axes. One finds that the minus sign in the invariant interval makes its presence felt, and "rotations" in the x,t plane look geometrically quite different from ordinary spatial rotations.

Clock Synchronization and Simultaneity

Allen has a watch with a small radio receiver that receives time signals from an atomic clock at the National Institute of Standards and Technology in Colorado. It saves him the nuisance of having to reset his watch occasionally. It is "synchronized," that is, always in agreement with, the national time standard. However, were Allen a zealot for precision, he would be bothered by the fact that his watch is always off by around a hundredth of a second, because that's the length of time it takes a radio signal to get two-thirds of the way across the United States to his watch in Massachusetts (a radio signal, like light, is one of Maxwell's electromagnetic waves and travels at the speed of light).

If Allen was really concerned about those hundredths of seconds, in principle, his watch could be exactly synchronized with the atomic clock at NIST by changing the watch's circuitry so that the reading of the watch took account of the time delay due to the transit time of the radio wave. Obviously, this is not really a serious problem for Allen. But it does illustrate what has to be included if you want to set up a frame of reference, at least conceptually: a series of clocks distributed throughout space, all of which show the same time. To do this, you can imagine taking a large number of identical clocks, along with radio receivers, and distributing them at strategic points throughout space in some inertial frame of reference, so that the clocks are all at rest relative to one another. You measure the spatial coordinates, $x, y,$ and z, of each of the clocks with the framework of meter sticks that constitutes the spatial part of the reference frame. The distance of the clock in question from the origin will then be r, where $r = \sqrt{x^2 + y^2 + z^2}$. You then send out a radio signal at a given time from the origin of the coordinate system, saying, "This is time $t = 0$." A person at each of the clocks then sets the clock to read, not $t = 0$, but $t = r/c$, to take account of the travel time of the light signal. You now have a set of clocks that are all at rest relative to one another and, as far as observers in that reference frame are concerned, all agree with one another.

Why did we put in that qualifying phrase, "as far as observers in that reference frame are concerned?" Let's go back to our reference frames S(earth) and S′(ship). Consider the time when the clock at the origin of S(earth) reads $t = 0$, and look at all of the clocks distributed along the x axis at various values of x in the earth frame. (The y and z coordinates don't get changed when you make a Lorentz transformation, so we'll just forget about them most of the time.) Since they are synchronized in the earth frame, they will all read t = 0. That is, the events corresponding to the hands of those clocks reading t = 0 appear simultaneous to observers in that frame.

What about for observers in the ship? The striking new feature of the Lorentz transformations is that the value of t' depends on both t and x.

Let's look at the clocks at the origin of the earth frame and at the point P on the x axis with x coordinate $x = x_1$. Consider two events: the event in which the hands of the clock at the origin in the earth frame read t = 0 and the event in which the hands of the clock at P in the earth frame read t = 0. The time and space coordinates (t,x) of the two events in S(earth) are thus (0,0) and (0,x_1), respectively, and they are simultaneous.

Now let's use the Lorentz transformation equations to find the time of the first event in S'(ship). Plugging x = 0 and t = 0 into the equation for t' gives t' = 0. No surprise there, but also nothing interesting since the convention we ad-opted was to take t = t' = 0 at the moment when the origins of the two reference frames passed one another, that is, when x = x' = 0. Notice, by the way, that at this moment the two clocks are momentarily at the same point, right next to each other. Observers in both reference frames can see them both simultane-ously and compare them unambiguously without having to send any signals back and forth.

But look what happens for the other event. Putting t = 0 and $x = x_1$ into the Lorentz transformation equation for t' gives $t' = \dfrac{-vx_1/c^2}{\sqrt{1-\left(v^2/c^2\right)}}$. Observers in the two reference frames do not agree on the time of the second event. More-over, observers in the earth frame think the two events were simultaneous, but those in the ship frame do not.

Why is this so? Before looking at the answer to that question, in order to avoid some possible confusion, let us take a moment to examine something the principles of relativity *do not* say, although on first reading you might be tempted to think that they do. They do not say that you will observe the speed of light to be c relative to every other object. They only say that will be true of objects that are at rest relative to you, that is to say, at rest in the inertial frame in which you are also at rest. For example, consider the following situation of a light pulse and a spaceship approaching one another, as observed in the earth's frame of reference. The light pulse is directed in the negative x direction and moves with speed c while the ship is traveling in the positive x direction with speed $c/2$. Then, after one second, provided the light pulse and the ship haven't actually met, the distance between them as measured by observers on earth, will have been *reduced* by 186,000 + 93,000 miles. Therefore, observers on earth will see the speed of the light pulse *relative to the ship* to be 279,000 miles per second, or $\frac{3}{2}c$. This does not violate the principles of relativity, because we are not in the ship's rest frame. We could carry out a transformation, using the correct (Lorentz) transformation equations, to a reference frame moving with speed $c/2$ in the positive x direction, that is, to the rest frame of the ship. The relative speed of the light pulse and the ship in that frame would be c, since the ship is at rest.

Thus, observers in different inertial frames do not agree on the relative velocity of two moving objects. In particular, the principles of relativity only guarantee that observers will measure the relative speed of a light pulse and an object to be c in the object's rest frame. We might note that this problem did not come up in the discussion of the Michelson-Morley experiment, since there we were dealing with the speed of two different light beams, moving in perpendicular directions, relative to the earth, as measured by observers on earth.

Let us return to the problem of why observers in the frames S(earth) and S'(ship) don't agree on the time at which the clock at the origin of S'(ship) passes the point P, where $x = x_1$. The problem is that observers in the two frames do not agree on the proper way to synchronize clocks. Observers in the earth frame synchronized the clock at P with the one at the origin by taking the travel time of a light signal in going to the right from the origin to P to be x_1/c, since they see the light moving at speed c relative to their frame S(earth). Observers in the ship frame also see light moving to the right relative to themselves at speed c, but they also see the earth moving to the left relative to them with speed v, since the ship is moving to the right relative to the earth. Thus, as we discussed in the previous two paragraphs, they will see the earth and the light moving toward one another, so to them the light is moving, *relative to the earth*, with speed $c + v$. As we discussed, observers in the two frames disagree on the relative speed of the earth and the light pulse. However, in agreement with the principles of relativity, both sets of observers see the light pulse moving with speed c relative to their own reference frame.

Observers in S'(ship) will thus say, "Those silly people in the earth frame don't even know how to synchronize their clocks correctly. They used c in computing the time delay due to the travel time of the light signal, when any fool can plainly see they should have used $c + v$. No wonder their clocks are incorrect and don't agree with the correctly synchronized clocks in our reference frame."

Observers on earth will, of course, have an equally dim view of observers in the ship frame as clock synchronizers for similar reasons. The fact that observers in each frame see light moving at speed c in their frame means that each set of observers uses a different, and *for them correct*, procedure to synchronize their clocks.

The Light Barrier

We mentioned in the introduction that special relativity is generally believed to rule out travel at speeds greater than the speed of light. A glance at the Lorentz transformation equations will indicate why this is so. You will see that, in transforming from the earth's reference frame to a reference frame moving with speed v relative to the earth, the equations for the coordinates in the new frame contain the expression $\dfrac{1}{\sqrt{1-(v^2/c^2)}}$. This may look a little complicated at first sight, but it's actually easy to understand. In an everyday situation, when v is much less than c, v^2/c^2 is very small and the denominator just becomes $\sqrt{1}$. So when v is small, this whole expression becomes a fancy way of writing "1." But as v gets close to c, the square root gets close to $\sqrt{(1-1)} = 0$. Since this factor is in the denominator, the overall expression gets bigger and bigger. Finally, if we tried to let $v = c$ exactly, then the denominator would be exactly zero. But division by zero is a mathematically meaningless operation whose result is undefined. So the Lorentz transformations are telling us that the relative velocity of any two inertial reference frames must be less than c. But the rest frame of a material particle moving at a uniform speed v would be an inertial frame. The fact that inertial frames are limited to speeds of $v < c$ thus seems to imply a similar limitation of the speeds of material particles. So the form of the Lorentz transformations implies the existence of a "light barrier" preventing matter from attaining the speed of light.

This conclusion also follows from the expression one obtains for the energy of a particle of mass m and speed v, if one imposes the condition that the laws of conservation of energy and conservation of momentum have the same form in all inertial frames, in consonance with the first principle of relativity. The resulting expression for the energy E of a particle of mass m and speed v is $E = \dfrac{mc^2}{\sqrt{1-v^2/c^2}}$. Because of the square root in the denominator, it would require infinite energy to accelerate such a particle to the speed of light, which is another way of saying that no material particle can ever actually attain the speed of light. If one looks at the derivation of this expression, which, though very pretty, is perhaps a bit too mathematical to give here, one sees that the square root in the denominator has its source in the corresponding square root in the Lorentz transformations. So again we see that the existence of a light barrier in special relativity arises from the requirement that the speed of light be the same in all inertial frames, which in turn leads to the Lorentz transformations.

"Massless" Particles and $E = mc^2$

Light, of course, does travel at the speed of light, a remarkably unremarkable statement. And in quantum theory, light does have a particle-like, as well as a wavelike, nature. A discussion of wave-particle duality would lead us much too far astray here. For our present purpose, we need just say that the "particles" of light, called "photons," or "light quanta" have $m = 0$. Hence, if we tried to apply the formula we've given for the energy of a material particle to light, we would find we had $0/0$, which is mathematically meaningless. However, it turns out that there is another way to write the expression for the relativistic energy of a particle in terms of its momentum and mass, which is given by $E^2 = p^2c^2 + m^2c^4$, where p is the momentum. This expression *does* make physical sense when $m = 0$. It says that the energy of a "massless" particle, such as a photon, has an energy, given in terms of its momentum, of $E = pc$, provided that the particle in question travels at the speed of light.[3]

The formula for the energy of a particle we've just given may not look very much like what you may have learned in an introductory physics course. There the discussion may have been confined to physics in the everyday, nonrelativistic limit in which v is very much less than c. In that limit, a standard mathematical result says that E is very well approximated by the formula $E = mc^2 + \frac{1}{2}mv^2$. The second term in this formula is the standard nonrelativistic expression for the "kinetic" energy of a particle, that is, the energy a particle has because it is in motion. In addition, the relativistic formula includes the famous new term mc^2, corresponding to a "rest" energy, which relativity predicts a particle has simply by virtue of its mass, even if it is at rest. Since that term is huge (c^2 is a very big number in ordinary units), you could reasonably ask why, if that term

---

3. This expression can be easily obtained from the relativistic expressions for momentum and energy: $p = \dfrac{mv}{\sqrt{1 - v^2/c^2}}$, $E = \dfrac{mc^2}{\sqrt{1 - v^2/c^2}}$.

is really there, no one noticed it before Einstein. The answer is that the rest energy, while huge, is also constant in the situations in which we ordinarily encounter it, because the number of particles of various kinds, with their associated rest energies, is constant. Such terms usually have no effect in solving the kinds of problems we are interested in, which involve the way things change with time or position, and constant terms generally occur on both sides of an equation and cancel out.

The rest energy can have dramatic effects in situations in which particle numbers do change with time. For example, a particle and its so-called antiparticle, a particle of the same mass but opposite electric charge (such as the positron in the case of the electron) can meet and annihilate one another, converting their entire combined rest energy into energy of electromagnetic radiation. Such phenomena had not yet been discovered experimentally in 1905 when Einstein published his paper on special relativity, and their later discovery provided powerful confirmation of the theory.

4
The Light Cone

Time past and time future
What might have been and what has been
Point to one end, which is always present.

    T. S. ELIOT, "Burnt Norton"

Well, the future for me is already a thing
of the past.

    BOB DYLAN, "Bye and Bye"

Absolute and Relative

Special relativity has shown us that time and space are different for different observers. A popular way in which one hears this expressed is with the phrase "everything is relative." But is that really so? For example, does the relativity of simultaneity imply that the causal order of events is also relative? By changing frames of reference, can we make the Second World War occur before Hitler's invasion of Poland? That is, can cause and effect be reversed by switching frames of reference? The world would be a pretty peculiar place if that were so.

We have seen that light has the same speed in all inertial frames. So the invariance of the speed of light is certainly not "relative," but is absolute in special relativity. This fact implies that Einstein's spacetime, unlike Newton's space and time, can be divided into regions, described by what is called the "light cone." In some of these regions, the order of events in time is the same in all frames of reference, and in others the temporal order is relative. As we will see, all pairs of events that are causally connected lie within a single region of the first type.

< 42 >

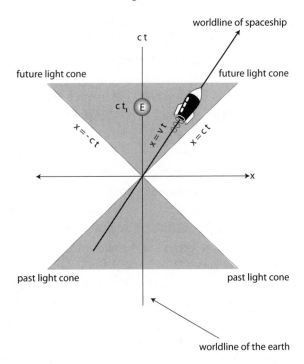

FIG. 4.1. The worldline of a spaceship moving with respect to the earth.

In figure 4.1, we present plots of the trajectory, that is, the variable ct versus the position x, for the earth and the spaceship in the reference frame S(earth). Such a trajectory is often called the "worldline" of an object. For ease and clarity of plotting, we confine ourselves to trajectories lying entirely in the two-dimensional x and ct portion of spacetime; all of the points we consider have $y = z = 0$. As we have done throughout, we choose the origin, $x = 0$, $ct = 0$, to be the point where the earth and ship pass one another, and the moment at which observers on both the earth and the ship choose to correspond to the origin of time on their respective clocks.

The earth is at rest at $x = 0$ in this reference frame; its worldline lies along the ct axis. Its position on the diagram at time t_1 is $x = 0$, $ct = ct_1$, ad we show a segment with t stretching from some distance in the past, where $t < 0$, and into the future, where $t > 0$. In order to plot the worldline of the ship we have arbitrarily chosen, $v = 0.8c$, so that the position of the ship is given by $x = 0.8ct$. The ship's trajectory on the ct versus x plot will then be a straight line with slope 0.8, relative to the vertical axis. All of our slopes are assumed to be measured with respect to the vertical axis (this is because, unlike the diagrams you are probably used to seeing, we have plotted ct along the vertical axis and x along the horizontal axis).

We choose the variable ct rather than t for convenience so that both axes have the same units of length. For an object moving with speed v, so that $x = vt$, the t versus x curve is a straight line of slope v. The cases of interest will involve values of v something like $c/2$, and since c is a huge number in normal units, the line in question would be almost vertical and indistinguishable from the x axis. By taking the variable ct, the slope of the ct versus x curve becomes a more manageable v/c. Using ct as the variable is equivalent to taking t as the variable, but in units of light-seconds (the distance traveled by light in one second, or about 300,000 kilometers), rather than seconds.

The two straight lines labeled $x = ct$ and $x = -ct$, in figure 4.1 describe the trajectories of light pulses moving in the positive and negative directions, respectively, along the x axis and passing through the origin at $t = 0$. These lines form what is called the "light cone." This is the portion of the spacetime surface $x^2 + y^2 + z^2 = (ct)^2$ lying in the ct versus x plane.

The light cone has a special significance, because it plays the same role in any inertial system. For example, as we know, if $x = ct$ in S(earth), it is also true that $x' = ct'$ in the reference frame S'(ship). The invariant interval s^2 between the origin and any point on the light cone satisfies the condition $s^2 = 0$, and, as we have discussed, s^2 is left unchanged if one makes a Lorentz transformation to a different inertial frame.

The light cone divides the page into four quadrants. The bottom and top portions of the light cone lie in the regions where t is, respectively, negative and positive. That is, they correspond to the regions of spacetime that are, respectively, before and after the time that we have called $t = 0$ when the spaceship passes the earth. To observers on earth and on the spaceship at $t = 0$, these regions are, respectively, in their past and future, and are called the past and future light cones. At points *inside* the past and future light cones, $x^2 - (ct)^2 < 0$, that is, s^2, the invariant interval between those points and the origin is negative. Such points are said to have a "timelike" separation from the origin. This is because the "time" part of the interval is larger than the "space" part.

Let's consider a particular event with time and space coordinates t_1 and x_1. If someone on earth at $t = 0$ wants to affect this event, then he must either travel or send a signal which travels at a speed u, where u is at least x_1/t_1. Because of the light barrier, we must have $u/c = x_1/ct_1 \leq 1$. That is, the slope of the ct_1

versus the x_1 curve cannot be greater than 1. Moreover, t_1 must be greater than zero, since we can only influence events in our future and not our past (we're not yet talking about time travel). These two conditions combine to describe the future light cone, so the future light cone is just the set of events which can be influenced by someone at the origin.

Let's look at some examples. Suppose that at $t = 0$, Starfleet Command on earth receives information that space pirates are planning to attack three space stations in exactly one year's time. The three stations are located at $x = 0.4$ light-years, $x = 1$ light-year, and $x = 1.2$ light-years. (A light-year, recall, is the distance light travels in one year.) This is before Starfleet has developed warp drive, so although they have spaceships with very powerful engines, they are limited by the light barrier. What will happen?

Refer to figure 4.2. The closest station is inside the forward light cone. Assuming that ships are available with top speeds greater than $0.4\,c$, one or more ships can be dispatched to support the station, and the ships will arrive before the marauding pirates. The second station is right on the edge of the light cone. No aid can reach it in time, since material objects cannot attain the speed of light. A signal can, however, be sent to the station using electromagnetic

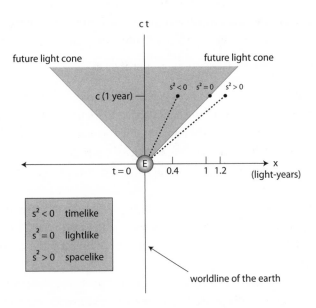

FIG. 4.2. "Space pirates." The figure depicts the difference between timelike, lightlike, and spacelike intervals.

waves, warning of the attack. (Unfortunately, it will not be a very timely warning, since it will arrive just as the pirates appear.)

The most distant station, which is outside the light cone, is out of luck. Nothing that Starfleet can do will influence events at that station one year in the future. Help cannot arrive before 1.2 years, and the station will have to fend for itself in the meantime.

Now let's consider the past light cone. The situation is similar, except the past light cone is the region of spacetime that can influence, rather than be influenced by, events on earth at t = 0. For example, worldlines of the earth and the ship stretch out of the past light cone to reach the origin, and past events on the ship, as well as the prior history of the earth, influence the earth at t = 0.

The interior of the light cone, where s^2 is negative, is thus the set of points which are in causal contact with the spacetime origin and can affect or be affected by what happens there. Since s^2 is invariant under the Lorentz transformations, the set of events in the interior of the light cone is the same in all inertial frames. The temporal order of events in the interior of the light cone, for example, whether an event is in the future or the past light cone of the event at the origin, is also a Lorentz invariant. (We will show this in the following paragraph.) Thus, given a pair of causally related events, observers in all inertial frames will agree as to which is the cause and which is the effect.

To see this, note that under a Lorentz transformation to an inertial frame moving with speed v relative to the frame S(earth), $t' = \dfrac{t - vx/c^2}{\sqrt{1 - \left(v^2/c^2 \right)}}$. Within the forward light cone, as we have seen, all values of x satisfy $x = ut$, where $u/c < 1$, and $v/c < 1$ because of the light barrier restriction on the speed of inertial frames. Now let's look at the numerator in t', which is $t - vx/c^2 = t - v(ut)/c^2$, where we have substituted $x = ut$. We can factor the right-hand side of this last equation to get $t - v(ut)/c^2 = t[1 - (u/c)(v/c)]$. Since u/c and v/c are both less than 1, their product is also less than 1. The denominator of t' is also always a positive expression. Thus t' involves t multiplied by a positive number, so that t' has the same sign as t. As a result, if an event at the origin causes a later event in *one* inertial frame, the effect will be seen to occur after the cause in *every* inertial frame.

On the other hand, events outside the light cone, even though they occur before t = 0, cannot influence the earth *at* t = 0, because the invariant interval $x^2 + y^2 + z^2 - (ct)^2 > 0$. (Points separated by such an invariant interval are said to have "spacelike" separation. This is because the "space" part of the interval is larger than the "time" part.) Suppose a Starfleet spy gained knowledge of the pirates' nefarious scheme two years in the past by overhearing some conversation in a bar on the planet Tatooine (for the purists, we know, we're mixing *Star Trek* and *Star Wars*) located 4 light-years from earth. The information is not going to do Starfleet any good at $t = 0$, since it can't reach them until $t = 2$ years, which is 1 year after the pirate attack occurs. The temporal order of events in the exterior of the light cone is not a Lorentz invariant and can be changed by a Lorentz transformation. (The argument in the preceding paragraph fails in this case because, for points outside the light cone, there is no guarantee that $u/c < 1$. However, the temporal order is not critical in this case, because, regardless of the sign of t, events outside the light cone can be of no help to Starfleet Command at $t = 0$.)

The Light Cone and Causality: A Summary

Because the light cone is so important for our future discussions, and since this is a rather difficult section, it's worth summarizing the ideas we've presented. We recommend a careful study of the following discussion and figure 4.3 (our treatment parallels that of Taylor and Wheeler, *Spacetime Physics*,

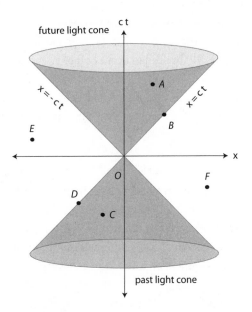

FIG. 4.3. The light cone. The event O represents the "present moment." The figure shows what events can affect, and be affected by, event O.

Sec. 6.3.[1] Figure 4.3 shows a light cone associated with an arbitrary spacetime event O (we have added one space dimension back in, to better illustrate the "cone").

Event A lies *inside* the *future* light cone of O, so O and A are separated by a *timelike* interval, for example, $s^2 < 0$. This means that a particle or signal traveling slower than light, emitted at O at t = 0, can affect what is *going to happen* at A. Event B lies *on* the *future* light cone of O, so O and B are separated by a *lightlike* ("null") interval, that is, $s^2 = 0$. Therefore, a light signal emitted at O can affect what is *going to happen* at B (in fact, the light ray arrives *just as* B occurs.) Event C lies *inside* the *past* light cone of O. This means that O and C are separated by a timelike interval, so a particle or slower-than-light signal emitted at event C can affect *what is happening* at O. Similarly, event D lies *on* the *past* light cone of O, so O and D are separated by a lightlike interval, and so a light signal emitted at D can affect *what is happening* at O. The events E and F lie *outside* both the past and future light cones of O, so each of these events are separated from O by a *spacelike* interval, that is, $s^2 > 0$. This means that for O to either affect, or be affected by, events E and F would require faster-than-light signaling. (A worldline connecting O with events E or F would have a slope of greater than 45°, and thus lie outside the light cone.) Therefore, events E and F can have *no* causal influence on O and vice versa. The time order of events A through D are invariant, that is, the same in all frames of reference. The time order of events E and F is different in different inertial frames. In some frames E and F will be seen as simultaneous; in other frames E will be seen to occur before F, or vice versa.

There is a light cone structure, like that depicted in figure 4.3, associated with *every* event in spacetime. The light cones define the "causal structure" of spacetime in that they determine which events can communicate with each other.

Note to the reader: Do not be disheartened if you did not understand everything in the last two chapters the first time through. They are probably the most demanding chapters in the book. You may need to read them more than once to fully grasp the ideas. However, an understanding of the concepts introduced here, *particularly* mastery of the notion of the "light cone," will be crucial to understanding the chapters on time travel and warp drives later on.

1. Taylor and Wheeler, *Spacetime Physics*, 2nd ed. (New York: W. H. Freeman and Co., 1992).

5

Forward Time Travel and the Twin "Paradox"

It was the best of times, it was the worst of times.

CHARLES DICKENS, *A Tale of Two Cities*

Baseball player: "What time is it?"
Yogi Berra: "You mean now?"

In the previous chapter, we saw that observers in two different inertial frames didn't agree on whether their clocks were synchronized initially. When observers in the frame of reference of the earth thought all their clocks read t = 0 at the same time, those in the spaceship frame disagreed, and vice versa. In this chapter we are going to see that observers in the two frames also disagree as to whether their clocks are running at the same speed. We will see that special relativity tells us that moving clocks appear to run slow. An observer who sees clocks in the other frame as moving though space will think those clocks are running slow compared to his own. Later in the chapter, we will see that this prediction leads to one of the clearest experimental verifications of special relativity and also to the conclusion that travel *forward* in time is possible.

Time Dilation and A Tale of Four Clocks

Recall that the two frames have coincident x and x' axes, with S'(ship) moving in the positive x (and x') directions with speed v relative to S(earth). Recall also that we placed clocks in the two frames at their respective origins and set them to read t = t' = 0 at the moment they pass one another. Let us call these two

< 49 >

The view from S(earth) at $t = t' = 0$

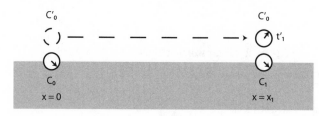

The view from S(earth) at $t = t_1 = x_1 / v$

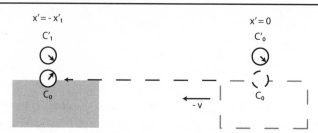

The view from S'(ship) at $t' = t'_1 = x'_1 / v$

FIG. 5.1. The time dilation effect.

clocks C_0 and C'_0, respectively. Since C_0 and C'_0 are momentarily side by side and simultaneously visible as they pass, observers in both frames will see them in agreement at that point. Observers in the earth frame will see C'_0 moving to the right with speed v along with the reference frame to which it is attached. Similarly, those in the ship will see C_0 moving to the left with speed v. Refer to figure 5.1 for the discussion in this section.

Now let's introduce a third clock into the mix, located in the frame S(earth) at the point $x = x_1$. We'll call this clock C_1. Since C'_0 is starting at the origin at $t = 0$ in the unprimed frame, and traveling with speed v, it will pass C_1 when C_1 reads t_1, where $x_1 = vt_1$. There is no relativity needed here. This statement involves three quantities all measured in the same reference frame, S(earth).

It's just the familiar formula that distance traveled equals speed multiplied by time, if all of these quantities are measured in the same reference frame.

Here, however, we do need some relativity. Now that we know *when* the clocks pass each other in the frame S(earth), we would like to know what C_0' *reads* when it passes C_1. That is to say, we have an event, C_0' passes C_1, whose coordinates in the frame S(earth) are $t = t_1$, $x = vt_1$. What is the time coordinate t_1' of that event as measured on the clock C_0', which is present at the event and at rest in S'(ship)? To answer that question, we need to use the Lorentz trans-

formation equation $t' = \dfrac{t - \left(vx/c^2\right)}{\sqrt{1 - \left(v^2/c^2\right)}}$ and put in the values for t_1 and vt_1 for

t and x, respectively. If we do this and factor out t_1, we get $t_1' = t_1 \dfrac{1 - v^2/c^2}{\sqrt{1 - v^2/c^2}}$.

Since for any quantity q, $q / \sqrt{q} = \sqrt{q}$, just from the definition of the square root, we arrive at the result

$$t_1' = t_1 \sqrt{1 - \left(v^2/c^2\right)} \quad .$$

Observers on earth agree that C_0' was set correctly at $t = 0$, because it agreed with their clock C_0 when the two passed each other. Now the time read on C_0' is less than that read on C_1, because the factor $\sqrt{1 - (v^2/c^2)}$ is smaller than 1 unless $v = 0$. Therefore, observers in the earth frame see the clock C_0', which for them is a clock moving with speed v, running slow compared to their clocks by a factor of $\sqrt{1 - (v^2/c^2)}$. Special relativity thus leads to the remarkable conclusion that *moving clocks run slow by the factor* $\sqrt{1 - (v^2/c^2)}$, compared to clocks at rest, where v is the speed of the clock. This phenomenon is called "time dilation."

There is a subtle point connected with this conclusion. Observers in both the S(earth) and S'(ship) frames see the two clocks C_1 and C_0' next to one another, and both agree that the time as shown on C_1 is greater than that shown on C_0'. Since C_1 is a moving clock in S'(ship), why don't observers in the ship frame come to the conclusion that moving clocks run *fast*? Observers on the ship agree that C_0 read correctly at $t' = 0$. However, C_1 is synchronized with C_0 according to observers in S(earth). As we discussed in the last chapter, observers in the two frames do not agree as to how to synchronize distant clocks. Therefore, observers on the ship say you cannot draw any valid conclusions from observations of C_1 because it wasn't set correctly to begin with.

The Lorentz transformations have been set up to guarantee that the principles of relativity prevail. This means that observers in any inertial frame must see moving clocks running slow, but they must determine this on the basis of experiments that are valid in their *own* frame. To allow observers in S'(ship) to do this, we must introduce a fourth clock, C'_1, which plays the roles in S'(ship) that C_1 played in S(earth). That is, C'_1 will be a clock at $x' = -x'_1$; the minus sign reflects the fact that C_0 will be moving in the negative x' direction relative to S'(ship). Remember now that C'_1 will be synchronized with C'_0 according to observers in the *ship frame*. If observers in S'(ship) compare the reading of what they see as the moving clock, C_0, with that of C'_1 as they pass, they will find that C_0, which was correct at $t' = 0$, is now reading slow.[1]

Note that the two events, clock C'_0 passing clock C_0, and C'_0 passing clock C_1, occur at the same place in S'(ship). Thus the time between these two events can be measured by a single clock, C'_0, in this frame. The time between two events that *occur at the same place in some inertial frame*, and thus can be measured by a *single* clock, is called the "proper time." In our example above, t' is therefore the proper time. The name is somewhat misleading, as it seems to denote the "correct" or "true" time. In fact, it implies neither of these. You can think of proper time as the time measured by your wristwatch as you travel along your worldline in spacetime.

To summarize, what we really mean by the phrase "moving clocks run slow," is that a clock moving at a constant velocity relative to an inertial frame containing synchronized clocks will be found to run slow when timed by these synchronized clocks. (An alternative, more geometric way of deriving the time dilation formula, without using the Lorentz transformations directly, can be obtained using a device known as a light clock, discussed in appendix 5.)

The Twin "Paradox"

In this section we will discuss one of the most famous "paradoxes" of relativity, the twin paradox. However, it should be noted at the outset that all of these standard so-called paradoxes of relativity, including the twin paradox, are really pseudo-paradoxes. That is, they only *seem* to be paradoxes because the principles of relativity have been applied incorrectly. This distinguishes

---

1. If you want to verify this, you will need what are called the inverse Lorentz transformation equations, which give t and x in terms of t' and x'. You can get these by taking the Lorentz transformation equations given in chapter 3, interchange t and t' and x and x', and replace v [the velocity of S'(ship) relative to S(earth)] with $-v$, since S(earth) will be moving to the left, in the negative x' direction, as seen from S'(ship).

them from the genuine logical consistency paradoxes which can occur in time travel, such as the grandfather paradox, which we will discuss at length in later chapters.

Let us introduce two twins, Jackie and Reggie, who are employed by a future space agency. Jackie is a crew member on a manned flyby of Alpha Centauri. The trip will be made using a rocket that will fly at constant speed to the star 4 light-years from earth, circle it, and return. (We ignore, very unrealistically, the periods of acceleration and deceleration at the beginning and end of the flight.) The rocket is capable of giving the spaceship a speed v, such that $1 / \sqrt{1 - \left(v^2 / c^2 \right)} = 10$. A little arithmetic will convince you that this means v is very nearly equal to c, the speed of light, so we will permit ourselves the luxury of saying (a space engineer surely would not) that the 8-light-year round-trip will require 8 years as seen by those on earth, though it would actually be a little longer. Jackie and Reggie are accustomed to reading a book every week; Reggie will read 416 books while Jackie is away, and Jackie stocks the spaceship library appropriately (with e-readers, naturally, to save weight).

Happily, the trip goes off without a hitch, and, 8 years later, Reggie meets the returning ship and the twins compare notes. Reggie is surprised to find that, for Jackie on the spaceship, only eight-tenths of a year have gone by, and the forty-second book is only about finished. Similarly, Jackie is surprised to find that, while less than a year has gone by on the ship, there are the results of two U.S. presidential elections to catch up with, and the campaign for a third is, alas, already well underway.

In short, while 8 years have gone by for Reggie and the rest of the outside world, less than a year has gone by for Jackie. This is just what we concluded in chapter 2 would constitute time travel into the future, and just what happens in the early pages of The Time Machine. Thus, we can say that Jackie has traveled more than 7 years into the future. The only difference is that Wells envisioned a time machine that remained stationary in space, while rapid travel through space is the mechanism that produces relativistic forward time travel. One could also achieve the same time dilation effect by traveling around a circular path within a relatively limited region of space, rather than out and back as with Jackie.

In the scenario we have discussed, there is no ambiguity as to which twin is younger, and therefore no ambiguity as to whose clock was running slow. The two twins are brought together again after the journey and can compare notes in person. Everyone agrees that it was Jackie, due to the time dilation on the moving spaceship, for whom time ran slowly.

But wait just a minute. The principle of relativity provides a sort of Declaration of Independence for inertial frames. It proclaims in ringing terms, "all inertial frames are created equal." Jackie sees the earth move away and then return. So one might conclude from this that Jackie would have read more books than Reggie. If this argument were true, one would conclude that special relativity indeed led to a paradox.

During the first half of the last century there was a fair amount of controversy engendered by this line of argument, with even some reputable physicists suggesting that it struck at the logical foundation of special relativity. In fact, there is no paradox, because there is a physical distinction between Jackie and Reggie. Reggie has remained at rest in the reference frame of the earth. Apart from corrections due to the earth's rotation and orbital motion, which are small because those velocities are very small compared to the speed of light, the earth is an inertial frame, moving with constant velocity. It is the reference frame that we have been denoting as S(earth). As an inertial frame, it is under the protection of the principle of relativity's grand proclamation of the equality of all inertial frames. The same is true of the frame S'(ship), since prior to the current discussion, we assumed that the ship was traveling with constant velocity.

This is not true, however, of the reference frame of Jackie's spaceship in the twin paradox. That frame cannot move with constant velocity, because if the twins are to be brought back together, the spaceship, traveling at relativistic speed, must reverse its direction and thus undergo acceleration. It is not under the protection of the principle of relativity's guarantee of the equality of inertial frames.

Invariant Interval and Proper Time

Consider the event, which we'll call E for convenience, in which a clock C located at $x = 0$ in a certain inertial frame, which we'll call S_E, reads $t = T$. Therefore the invariant interval between E and the spacetime origin O (with coordinates $(0,0)$) is $s^2 = -(ct)^2 + (x)^2 = -(cT)^2$. The elapsed time on the clock, which was present at both the spacetime origin O and E, is thus $\sqrt{-s^2 / c^2}$. (Recall that for timelike intervals, $s^2 < 0$, so $-s^2 > 0$, and the quantity under the square root is therefore positive.) The time of an event on a clock present at both the spacetime origin and the event E, is the proper time of the event, as measured along that clock's particular worldline. However, proper time is not unique, in the sense that it depends on worldline of the clock in question between the origin and E. Here we have the special case that the clock is at rest in an inertial frame, and in that case we have the simple relation given above between the proper time on that clock and the invariant interval (this gives us a new additional property of the invariant interval, which we didn't know before).

Now let's say that, instead of a clock C remaining at rest, we consider a clock C' that goes from the origin O with coordinates $(0,0)$ to E with coordinates $(0, cT)$ by first moving at constant velocity v to the intermediate spacetime event A, with coordinates $(x, ct) = \left(\dfrac{vT}{2}, \dfrac{cT}{2} \right)$. It then travels from event A to event E along another path of constant speed v, but in the opposite direction. That is, we first give the clock a "kick" in the positive x direction, and then a kick in the opposite direction. In the twin paradox, C' would correspond to a clock on the spaceship, in the approximation in which we assume that Jackie's ship flies to Alpha Centauri at constant speed and then turns immediately around and flies back, neglecting all the speeding up and slowing down along the way. This is shown in figure 5.2. The heavy black lines represent

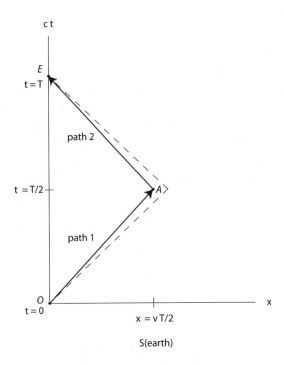

FIG. 5.2. The twin paradox. Reggie's worldline is the straight line connecting events O and E. Jackie, the spaceship twin, follows the "bent" worldline OAE. In this figure, Jackie's acceleration and deceleration periods are ignored.

the two legs of Jackie's trip, outbound and then return. The dotted lines represent the paths of light rays. The fact that the solid lines are so close to the dotted lines indicates that Jackie's spaceship is traveling very close to the speed of light.

We already calculated the proper time elapsed along a straight worldline from O to E, which would correspond to the time elapsed for the stay-at-home twin, Reggie. That was simply T. Let us now calculate the proper time elapsed along a "bent" worldline for Jackie. In this case, we can't find the invariant interval, and hence, the elapsed proper time on the clock C' at one stroke, because the directions of travel along the two segments of the path are different. But since the clock moves at the same constant speed along each side, we can use the invariant interval for each side to find the elapsed time for each segment of the trip, and since elapsed time has no direction, they can be added to get the total elapsed time.

The invariance of the spacetime interval can be expressed as

$$s^2 = -(ct')^2 + (x')^2 = -(ct)^2 + (x)^2$$

where t' will denote the proper time along the "bent" worldline. This would be Jackie's "wristwatch time." Let us first calculate the proper time elapsed along the first leg of Jackie's trip, from O to A. We'll call this part of the trip path 1 and call the spacetime interval along this path s_1^2. In Jackie's frame, all events occur at the same place, namely, $x' = 0$. Therefore the spacetime interval, in terms of her coordinates, is just $s_1^2 = -(ct'_1)^2$, where t'_1 is the elapsed proper time for Jackie along path 1.

To get the spacetime interval along path 1 in terms of Reggie's coordinates, notice that the coordinates of event A in S(earth) are $x = vT/2$, $ct = cT/2$. From the invariance of the spacetime interval, all observers must agree on the value of the interval *along a given path*. Therefore, our earlier equation becomes

$$s_1^2 = -(ct'_1)^2 = -(cT/2)^2 + (vT/2)^2.$$

Multiplying both sides by −1 and factoring out c^2 and $(T/2)^2$ gives

$$c^2(t'_1)^2 = \frac{c^2 T^2}{4}\left(1 - \frac{v^2}{c^2}\right).$$

If we now cancel the c^2, and take the square root of both sides, we get

$$t'_1 = \frac{T}{2}\sqrt{1 - \frac{v^2}{c^2}} \; .$$

Since the bent path is symmetrical, it's not too hard to convince yourself that the spacetime interval along path 2 will be equal to that along path 1, that is, $s_1^2 = s_2^2$. An identical calculation to the one just performed would then show that the elapsed proper time along path 2, t_2', is the same as that along path 1. Hence, the total proper time along the bent path is given by $t' = t_1' + t_2'$, and the elapsed proper time for Jackie for the entire trip is

$$t' = T\sqrt{1 - \frac{v^2}{c^2}} \; .$$

Therefore, the unambiguous result is that $t' < T$, which means less time has elapsed for Jackie than has for Reggie. So Jackie is the younger of the twins when they reunite.[2]

You might be worried about the fact that we ignored the periods of acceleration and deceleration during the trip. To show that this is not crucial to the argument, let's look at figure 5.3, where we have "rounded off the corners" of Jackie's trajectory to include these effects. We could, if we wished, break up the curved path into a lot of tiny (approximately) straight-line segments. Then we could work out the proper time along each straight-line segment, as we did in the previous example, and add them up. Our result would still be that Jackie is younger than Reggie when they reunite. This also dispels the commonly cited fallacy that because acceleration is involved, special relativity is not applicable and one must use general relativity to resolve the paradox.

In both figures 5.2 and 5.3, the "bent" line path between O and E is actually shorter, in terms of elapsed proper time, than the vertical straight-line path between the same two events! But, you say, it certainly doesn't look that way in the figure. This is because we are forced to illustrate the geometry of spacetime in special relativity (you need to remember the minus sign in the interval!) on a piece of paper which has the geometry of Euclidean space. It may help you to recall that, in our earlier spacetime diagrams, lines inclined at 45° (representing the paths of light rays) actually have zero length in spacetime, that is, $s^2 = 0$. In figure 5.2, the two legs of Jackie's trip lie very close to 45° lines, and hence, together have much shorter length (in terms of proper time) than the vertical straight-line path.

2. A good fictional portrayal of a forward time travel scenario is given in Poul Anderson's novel *Tau Zero*, Gollancz SF collector's edition (London: Gollancz, 1970).

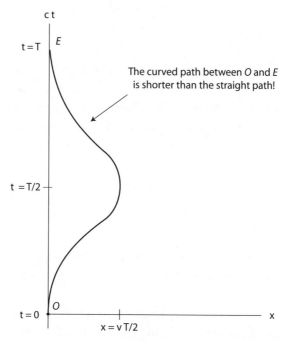

The curved path between *O* and *E* is shorter than the straight path!

FIG. 5.3. A worldline for a spaceship observer including periods of acceleration and deceleration. These make no difference to the main argument. As a result of the geometry of spacetime, the curved worldline connecting events and is actually shorter, in terms of proper time elapsed, than the straight worldline connecting the same two events!

Practical Considerations and Experiments

Special relativity clearly allows the theoretical possibility of traveling forward in time. In this section we will examine briefly why such trips are not a very realistic possibility for human beings or other macroscopic objects, as well as the evidence that they are rather commonplace in the world of elementary particles.

Suppose you really cannot wait to see what kind of electronic miracles await us 20 years in the future, and you're only patient enough to spend 2 years getting there. In order to make the trip, you need a space capsule large enough to accommodate you that is capable of achieving a speed v through space, such that $\sqrt{1-\left(v^2/c^2\right)} = 1/10$. Then the clocks on the ship, including your own biological clock, run at about one-tenth the rate of clocks outside. Since the energy of an object is $mc^2/\sqrt{1-(v^2/c^2)}$, you would need to increase the total energy of your space capsule by about 9 mc^2 to bring its speed up to v. How much energy this is depends, of course, on m. Let's try a value of about 1,000 kilograms for m. That's about the mass of a car, and your capsule would no doubt have to be much larger. But even for a mass of 1,000 kilograms, it turns out that mc^2 is about equal to the entire annual electrical power output of the United States! Thus, all the generators in the country would have to devote their full time for a year to supplying power for your projected trip. There are other serious technical problems as well. But the energy requirement by itself is enough to demonstrate that time travel into the future using relativistic time dilation is not going to be practical anytime in the near future, if ever.

One experiment has been done in which a macroscopic object was sent into the future. The object was an atomic clock, and it was flown around the world on a commercial jetliner. The typical speed of such planes is around 500 miles per hour, or around $\frac{1}{7}$ of a mile per second, which gives a value of v^2 / c^2 of less than 10^{-12}. Nevertheless, the experimental group reported that the clock on the plane lost about 1 nanosecond (one-billionth of a second) during the trip, compared to a corresponding clock that had remained behind on the ground. That was about the limit of the accuracy of the experiment. A supporter of special relativity would not feel terribly secure if the experimental evidence in support of time dilation hung only by this rather slender nanosecond. (Actually, the result of this experiment is a test of *both* Einstein's special theory of relativity and his general theory of relativity, that is, his theory of gravity. The calculations done by the experimenters to make their prediction depended on both the aircraft's speed [special relativity] and the height of the aircraft above the earth's surface [general relativity].)

Fortunately, there is a wealth of other evidence from the world of elementary particles. Physicists at high-energy labs routinely observe these small masses achieve relativistic speeds; we also observe such speeds for cosmic ray particles incident on earth from outer space. Many of these particles are radioactively unstable and decay with a well-established time interval called a "half-life." That is, half the particles decay, on the average, after one half-life, half of the remainder after the next, and so on, and the rate of decay can be observed by detecting the decay products with various detectors such as Geiger counters or photomultiplier tubes. As a result, a sample of a number of such particles provides a clock. It is routinely observed that particles produced with higher energy—and thus with speeds close to the speed of light, and hence, a smaller value of $\sqrt{1 - v^2 / c^2}$ —decay more slowly. That is, they have a longer half-life, as seen in the laboratory, than similar particles that decay at rest. In general the observations are consistent with the relativistic prediction that the time read on a moving clock, which is inversely proportional to the half-life in the case of decaying particles, is proportional to $\sqrt{1-v^2/c^2}$, or equivalently, to mc^2/E.

One experiment of particular interest involved a circulating beam of particles called muons. These particles decay radioactively with a half-life of about a microsecond. The main purpose of the experiment was to compare magnetic properties of muons and electrons. In the process, it was confirmed with rather high precision, that the lifetime of muons in motion was equal to the known muon half-life at rest multiplied by the predicted factor of $1/\sqrt{1-v^2/c^2}$. In contrast to most such experiments, which involve linear beams of particles, this one involved a circulating beam with the circulating particles returning periodically to their starting point. Thus, it modeled the twin paradox. To no one's surprise the circulating muons, playing the role of the traveling twin, underwent time dilation, compared to muons remaining at rest.

The predictions of special relativity are tested literally thousands of times a day in high-energy physics accelerators all around the world. In fact, the "nuts and bolts" engineers who design these accelerators must take into account the effects of special relativity, such as the increase in energy with velocity. Otherwise, their machines would not function.

A Final Look at Forward Time Travel through The Door into Summer

We'll conclude this chapter with a quick look at another possible mechanism for forward time travel—one that does not involve relativity nor primarily even physics, but rather, biology and medicine. The look is inspired by Robert Heinlein's book, *The Door into Summer*. If we had the time, we would be able to meet one of the most engaging groups of characters, both human and feline, in science fiction. However, we must forego such pleasures and attend to business.

In the book the protagonist travels forward, then back, and then forward again in time. Not surprisingly, Heinlein does not provide a detailed mechanism for backward time travel, resorting instead to a glorified "black box." But he does provide a mechanism for the forward time travel parts of the journey, namely, "cold" or cryogenic sleep. The characters' bodies are cooled to liquid helium temperatures, after which it is hypothesized that all aging processes stop. That is, the biological clocks of those stored are slowed—essentially stopped—with respect to the flow of time in the outside world until they are brought out of storage at some prearranged future time.

When Allen proposed this to his time travel classes as being a form of time travel, his students tended to rebel and think he was cheating. It clearly wasn't what they were used to thinking of as time travel, perhaps because the travelers were all too clearly there throughout the process, rather than in an invisible time machine (which, in fact, should have been visible if Wells had gotten the physics right). Actually, Heinlein's scheme is exactly the sort of thing we said in chapter 2 would constitute forward time travel, namely, time going by very slowly for a time traveler inside a time machine relative to the rate at which it was going by outside.

In this area it seems likely the necessary physics has already been done. Although low-temperature physicists continue to make advances toward the unreachable goal of absolute zero, they are already so close that the progress comes in small fractions of a degree that seem unlikely to be relevant to the cold sleep problem. One guesses that, if this sort of forward time travel can be done at all, it can be done at the very low temperatures already attainable.

We are neither MDs nor trained biologists and have no wisdom to offer on the likelihood, or even the plausibility, of cryogenic time travel ever becoming a reality. It's not clear to us whether practitioners of the relevant disciplines are in a position at this stage to offer any wisdom either. However, given the technological problems confronting the relativistic version, which we've only touched on, it's not impossible to imagine that forward time travel will turn out to be a field for the biologists, not the physicists.

In the meantime, if you haven't read it and you come across a copy of The Door into Summer, get it; it's a fun read.

6

"Forward, into the Past"

Fritz Fassbender: "I decided to follow you here."
Michael James: "If you followed me here, how did
    you contrive to be here before me?"
Fritz Fassbender: "Eh, I followed you . . . very fast."

What's New, Pussycat?

The beginnings of Allen's participation in the writing of this book can be traced to a specific time and place—in relativistic lingo, to a specific event. The time was midwinter 1967. The place was the reading room of the library at the Lawrence Berkeley National Laboratory, a high-energy physics research laboratory operated for the Department of Energy by the University of California at Berkeley. Allen was in the middle of a streak of extraordinary good fortune. He was currently enjoying his first sabbatical leave, one of the perks of his recent promotion to tenured rank at Tufts University. Following the completion of his graduate work at Harvard, he had accepted an appointment at Tufts, which was in the town of Medford, adjacent to Cambridge, and was just beginning to develop a PhD program in physics. Over the years this proved a felicitous decision in a number of ways. The most important of these was that, just before her graduation in 1964, he met a strikingly pretty Tufts senior, Marylee Sticklin, who was about to receive her degree in biology summa cum laude and, happily, begin studying for her own PhD in plant physiology at Harvard, conveniently nearby.

Things became a little less convenient a year later when her thesis advisor left Harvard to accept a position as provost of one of the colleges at the new branch of the University of California, which was about to open at Santa Cruz. He invited Marylee to transfer to Santa Cruz so she could continue working

< 62 >

under his supervision at his new post. But by that time, a mere 3,000 miles was not going to stand in the way. Allen and Marylee became engaged during his spring vacation visit to Santa Cruz in 1966 and were married in July of that year. By living in San Jose and both enduring rather tedious commutes, Allen was able to spend his sabbatical visiting the research group at Berkeley led by professors Geoffrey Chew and Stanley Mandelstam, at the time one of the most exciting places a young theoretical elementary particle physicist could find himself. Marylee, meanwhile, could get on with finishing her dissertation at the beautiful Santa Cruz campus, looking out over the Pacific from amidst groves of redwood trees. Thus Allen's sabbatical was also a kind of yearlong honeymoon, which was a fitting beginning for an idyllic marriage of some 42-plus years.

Superluminal Particles!?

Now that we know how he got there, however, let's get back to that seminal event for Allen in the Berkeley lab library. On the morning in question he was glancing through the latest stack of preprints when he came across a paper by Professor Gerald Feinberg of Columbia University. (In those days, physicists often sent out to colleagues as well as to libraries advance copies, or "preprints" of their current papers that had not yet appeared in print in a professional journal. Nowadays, we post our papers to what's called the "e-print archive," where the next day, they become freely available to anyone in the world. If you are interested, the URL is http://xxx.lanl.gov (contrary to what you might think, this is *not* the Department of Homeland Pornography).

Feinberg noted that what was really prohibited by special relativity was not actually travel faster than the speed of light, but rather, the acceleration of ordinary matter to such speeds by going *through* the speed of light, where the Lorentz transformations become meaningless and the energy of ordinary particles with mass becomes infinite. What, Feinberg imagined, if there was a class of particles that *always* moved at speeds greater than c? He proposed the name "tachyon," based on the word in ancient Greek for "swift," for such particles, and suggested that their energy in terms of their speed v might be given by the expression $E_t = \dfrac{m_t c^2}{\sqrt{v^2/c^2 - 1}}$, where $v > c$. Therefore, we have the square root of a positive number in the denominator, and we avoid the occurrence of the "imaginary" number $i = \sqrt{-1}$, which you probably encountered in high school algebra. Imaginary numbers can be very useful in various roles in both mathematics and physics. However, there is no actual, or "real," number that, when squared, gives −1. Therefore, physical quantities, which can be observed and measured with an instrument such as a clock or scale, must always be given in real numbers.

Notice from the equation that the energy of a tachyon becomes infinite as its velocity *decreases* toward the speed of light. This is analogous to the behavior of ordinary particles whose energy becomes infinite as their speed *increases* toward c. Thus, just as the speed of ordinary particles is confined by the light barrier to be *subluminal*, that is, less than the speed of light, tachyons, if they exist, would be confined to *always* travel at superluminal speeds, that is, at speeds u where $u > c$. Also notice that, in contrast with ordinary particles, and contrary to one's intuition, the energy of a tachyon would *decrease* as its speed increases and would actually become zero as the speed becomes infinitely large.

Feinberg was actually not the first to come up with this general idea. O. Bilaniuk, N. Deshpande, and E.C.G. Sudarshan had proposed an idea similar to Feinberg's, though differing in some important technical details, about three years earlier. Their article was published in a journal that was less frequently read by research physicists than the *Physical Review*, in which Feinberg's paper had appeared. As a result, their article had attracted somewhat less widespread attention. Also, they had not supplied a name for their suggested new particle. Sometimes, even in the world of physics with its strict standards of scholarship, a catchy name for a particle can help draw attention to a new idea.

Tachyons and Paradoxes

Allen thought the tachyon idea was clever and interesting. However, as was pointed out in Feinberg's paper, it had a major problem in the shape of potential paradoxes—real paradoxes this time—that could not be walled off behind quotation marks as in the title of the previous chapter. The paradoxes arise because tachyons, by definition, have speeds greater than the speed of light, so their worldlines are spacelike. Therefore, as discussed in chapter 4, the temporal order of events along their worldlines is not the same in all inertial frames. This means that tachyons threaten the same kind of paradoxes as those associated with backward time travel. The existence of tachyons would not allow people, made of ordinary matter, to travel at superluminal speeds. But it does raise the possibility of using tachyons to send information at speeds $u > c$. In relativity, this in turn leads to the possibility of sending information backward in time, and this can lead to potentially paradoxical results similar to those encountered in science fiction stories involving backward time travel by humans.

To see how this could happen, let us suppose that observers on earth have a device that allows them to produce tachyons of speed $u > c$. Let us suppose they produce such a tachyon at time $t = 0$, traveling in what we will define to be the positive x direction relative to the observers' rest frame, S(earth). The tachyons are later detected, after a time $t = t_d$, at a point with coordinate $x = x_d = ut_d$ in S(earth). Since $u > c$, the spacetime point (t_d, x_d) lies outside the light cone, and therefore, as we discussed in chapter 4, the sign of t_d is not the same in all inertial frames.

To observers on the spaceship, moving with speed v in the positive x direction in S(earth), the tachyon will be detected at time $t' = t_d'$ and position $x' = x_d'$ (as usual, we set the clocks at the origins of the two reference frames to read $t = t' = 0$ as they pass one another). We can use the Lorentz transformations to find that, in their reference frame S'(ship), observers on the ship will see the tachyon detected at

$$x_d' = \frac{1}{\sqrt{1-(v^2/c^2)}}(x_d - vt_d)$$

$$t_d' = \frac{1}{\sqrt{1-(v^2/c^2)}}(t_d - (vx_d/c^2))$$

If we substitute $x_d = ut_d$ into the second of these two equations and factor out t_d, we obtain

$$t_d' = \frac{t_d}{\sqrt{1-(v^2/c^2)}}(1-(vu/c^2))$$

Remember that in these equations, v is the speed of the ship and is subluminal, that is, $u < c$. From this last equation we see that if $u > c^2/v$, $t_d' < 0$. That is, if it is possible to generate a fast-enough tachyon, it will be sent backward in time in the ship's inertial rest frame S'(ship) and be detected before it is produced, according to observers on the ship. Even if u isn't big enough to satisfy the condition with v equal to the ship's speed, one can always find some inertial frame whose speed is high enough, but still less than c in which the tachyon will travel backward in time.

The mere fact is that $t_d' < 0$ doesn't by itself open the way to possible paradoxes. That only happens if we can send a return signal from the event at which the tachyon was detected, which reaches $x = 0$ before $t = 0$. In that case, we could arrange for the receipt of the return signal to block the transmission of the original tachyon from the origin and we would then have a paradox in which the tachyon is sent if—and only if—it isn't sent.

Since the event at which the tachyon was detected occurred outside the light cone of the origin, the return signal would have to exceed the speed of light and thus involve a second tachyon in order for the paradox to arise. In the ship's reference frame, the second tachyon would travel forward in time, covering the spatial distance of length x_d' back to the origin in a positive time less than $-t_d'$ (remember $t_d' < 0$). Hence, the speed of the return tachyon must satisfy $u_r > x_d' / (-t_d')$, which, after some algebra, one can show implies that $u_r > u$; that is, the tachyon used for the return signal must be somewhat faster than the original tachyon. The principles of relativity guarantee that if it was possible for observers in S(earth) to build a device that would produce tachyons traveling forward in time in their rest frame, then it must also be possible for observers in S'(ship) to produce such tachyons in their rest frame, and thus send the return signal. Therefore, the existence of tachyons, coupled with special relativity, seems to result in paradoxes.

The Reinterpretation Principle

Sudarshan and his colleagues suggested a possible way around the problem, which they named the "reinterpretation principle." To understand this, we must first take a moment to consider some implications of special relativity for the energies of tachyons. For ordinary particles, it can be shown that the Lorentz transformations imply that the sign of the energy of a particle, just like the sign of the temporal order of two points on the worldline, is the same in all inertial frames. Therefore, all observers will see the particle as having positive energy, though they will disagree on how much energy it has. However, for a tachyon of energy E in the earth's frame, it turns out that its energy E' in the frame S'(ship) is given by $E' = \left(\dfrac{E}{\sqrt{1 - v^2 / c^2}} \right) (1 - (vu / c^2))$. A glance at the equation for t_d' reveals that E' becomes negative when t_d' does. Thus, a tachyon that travels backward in time always has negative energy.

With this in mind, Sudarshan asked what would actually be seen by an observer "living forward in time" in S'(ship) when the tachyon is detected at t_d'. At that point the tachyon detector absorbs a tachyon from the future carrying negative energy. But absorbing negative energy is the same thing as losing positive energy. (Incurring a charge of \$1,000 on your credit card has the same effect on your net worth as taking \$1,000 out of your checking account. In either case you are poorer by \$1,000.) Thus, it will appear that the detector has lost energy by emitting a positive-energy tachyon that appears at t'_d and continues to be present, appearing to be moving forward in time along with the observer. This will continue until $t' = t = 0$. This is the time at which the tachyon was originally emitted backward in time with negative energy E' by the tachyon production device, as seen in S'(ship). At that point the tachyon will seem to an observer living forward in time in S'(ship) to disappear into the production device. The observer will thus conclude that the device, rather than having emitted a negative-energy tachyon traveling backward in time, has absorbed a positive-energy tachyon (actually it would be an anti-tachyon, the antiparticle of the tachyon, but we will skip over this technical point) traveling forward in time and thus coming from the past. These two processes again have the same physical effect. The absorption of positive energy or the emission of negative energy both lead to a gain in energy, just as either depositing a check or paying of a credit card balance of the same size increase your net worth by the same amount.

Sudarshan argued that observers in the ship would not recognize that they had received a message from the future. They would instead "reinterpret" the occurrence as the spontaneous emission of a tachyon from their detector. They would not recognize that they had received information from the future and thus would be unable to act on it to produce a paradox.

The difficulty with this analysis was pointed out rather quickly in two articles, one by W. B. Rolnick and the other by G. Benford,[1] D. L. Book, and W. A. Newcomb. Both articles agreed that the reinterpretation principle would allow one to avoid paradoxical consequences in cases where only a *single* tachyon was involved. However, by controlling the tachyon transmitter, one could send an extended message, say by Morse code, spelling out, "To be or not to be, that is the question." While it might look to observers on the ship as though their transmitter was producing this at random, they would soon recognize that they were seeing an intelligible message. The odds against that happening by chance are astronomical, and thus they would conclude that someone had sent them a message. (You sometimes read something to the effect that if you put

---

1. Professor Gregory Benford, who is a member of the faculty at the University of California, Irvine, has written a number of excellent "hard" science fiction novels. An early one, *Timescape*, deals with using tachyons to warn the past about an impending ecological disaster. Although an excellent book, the fact that it is set in what is now the past may be somewhat jarring for contemporary readers, who may be more accustomed to science fiction set in the future.

a monkey at a typewriter and let it type randomly, it will eventually reproduce all the books in the British Museum. While there is a certain abstract sense in which this could be considered correct, it is so far from any practical significance as to be basically meaningless. It would, in fact, take a monkey many times the accepted age of the universe since the big bang to reproduce a single page of this book.) Hence, the reinterpretation principle would not eliminate the possibility of communication with the past if tachyons existed.

A Problem with Superluminal Reference Frames

Given that tachyons seemed to lead to paradoxes, which were unacceptable and also unavoidable, Allen was inclined to give up his brief interest in the idea. However, he was collaborating actively on elementary particle physics projects with Adel Antippa, who had recently completed his PhD at Tufts under Allen's supervision and was now on the faculty at the Université du Québec à Trois Rivières. Antippa had been bitten by the tachyon bug and was eager that he and Allen should also undertake a collaboration in this field.

After some persuasion, Allen agreed to join in examining whether some further developments might be made on the basis of a paper by Leonard Parker, a noted expert in the general theory of relativity, as Allen was to learn later, at the University of Wisconsin–Milwaukee. To understand what Parker had done, we should first look at what had *not* yet been done. While tachyons were, by assumption, particles whose speeds exceeded the speed of light, the class of allowed inertial frames continued to be limited to those with subluminal velocities relative to one another.

On the other hand, if tachyons existed, it is at least conceivable that they could be used to make clocks and meter sticks from which, in turn, reference frames could be constructed. Such reference frames, like their constituent particles, would presumably be superluminal, with speeds $v > c$ relative to subluminal reference frames. One would then need some generalization of the Lorentz transformations to relate the coordinates of events in superluminal and subluminal reference frames. Hopefully these would be such as to leave the speed of light invariant in going from one class of reference frames to the other, so that some kind of extended principles of relativity would exist.

Parker showed how to construct such a theory very neatly in a kind of "toy" spacetime that had, as usual, a time dimension, but only one space axis, let's say an x axis. Sometimes, studying such toy spaces can give insight into the actual four-dimensional problem of interest. In the two-dimensional case, consider a superluminal frame with constant speed $v > c$ relative to a subluminal frame. Parker's transformation equations giving the coordinates of an event (t',x') in the superluminal frame in terms of the coordinates (t,x) in a subluminal frame were simply the Lorentz transformation equations with the time and space axes interchanged in the superluminal frame. Thus, instead of having $(ct')^2 - x'^2 = (ct)^2 - x^2$, as for a subluminal transformation, the superluminal transformation gave $-(ct')^2 + x'^2 = (ct)^2 - x^2$. For the case of a light ray traveling in the x direction, when $(ct)^2 - x^2 = 0$, the minus sign made no difference and the sanctity of the speed of light was preserved in both the subluminal and superluminal frames.

Antippa and Allen noticed that while, as usual, the temporal order of events along the worldline of a tachyon was not the same in all inertial frames, the spatial order along the x (and x') axis was. One could therefore consistently postulate that tachyons moved only in the positive x direction, just as ordinary particles move only in the positive time direction. This would rule out paradoxes, since neither ordinary particles nor tachyons could return to both a time *and* position, that is, to an event, at which they had already been present, a necessary condition for creating a paradox.

All of this was very nice, but unfortunately only applied to a make-believe world with only one space dimension. Antippa and Allen did construct a four-dimensional world with these features, but it was a very ugly world indeed. It had a preferred direction picked out, namely, the one along which the superluminal transformations were allowed. The trouble was that, in the real four-dimensional world, there were three directions in space and only one in time, so two of the spatial axes were left without a temporal partner with which they could be interchanged. (There was a published proposal for a theory with three different time directions. Allen actually spent a couple of days thinking about how you might give a physical interpretation to the other two time directions and then threw up his hands in surrender!)

A preferred direction in space was anathema in physics; it was like singling out a preferred inertial frame in special relativity, only worse. It is intuitively natural to assume that any direction in space is as good as any other. You might argue that down is clearly different from sideways or up, but that's only because of what is, in the grand scheme of things, a mere coincidence. The distinction between down and sideways is not telling us anything fundamental about the laws of physics, but only about the particular place in which we find ourselves. It just happens that we live where there is a modest-sized astronomical body, the earth, in the "down" direction. And for that matter, of course, the down direction in space for us is the up direction for our good mates in Australia, a mere 12,000 miles away on the opposite side of the earth. And 12,000 miles is pretty darn "mere" on the scale of the universe.

So it is intuitively simple and natural to think that nothing in the laws of physics picks out a particular direction in space as preferred. In physics, this is called the assumption that space is "isotropic." Equivalently, we say that space is invariant, or symmetric, under rotations of the coordinate axes. We've already made use of this assumption a number of times in this book without stopping to think about it. We've assumed repeatedly, without really giving it any thought, that we could choose our coordinate system in a particular situation so that the x axis pointed in some particularly convenient direction.

In fact, not only does it seem simple and natural to assume that space is symmetric under rotations, there is abundant and extremely powerful experimental evidence that this is the case. One of the most beautiful themes in theoretical physics, which recurs in many contexts, is the connection between symmetries exhibited by the laws of physics, and the existence of conservation laws that can be derived just from those symmetries. One of the most fundamental conservation laws, called the conservation of angular momentum, is a consequence of—and direct evidence for—the isotropy of space. It is not as well known as its more famous brethren, conservation of linear momentum and conservation of energy (which also follow from symmetries). What, if anything, you may have heard about angular momentum will depend on your physics course background. However, conservation of angular momentum has applications that are equally widespread, and its validity is supported by exquisitely precise measurements on the behavior of atomic nuclei.

So a theory of tachyons that involved a preferred direction was not terribly appealing aesthetically and could be viable experimentally only if the coupling of tachyons to ordinary matter was extremely weak, so the resulting violations of the law of conservation of angular momentum would be too small to be observed. Nevertheless, the idea that, if there were tachyons, there should be superluminal Lorentz transformations, seemed natural and prompted a good deal of discussion in the physics literature. Allen and Antippa did some additional work on the model; in particular, they wrote a paper working out the form Maxwell's equations for charged tachyons would take after a superluminal coordinate transformation. They were joined in this endeavor by Louis Marchildon, a very capable student of Antippa. This gave Allen a chance to enjoy collaborating not only with a former student but with, so to speak, an academic grandchild. The most useful aspect of this paper was probably that it corrected a rash of articles that had appeared in European journals claiming it was possible to construct a theory of superluminal coordinate transformations that did not involve the introduction of a preferred direction. Allen and his collaborators were able to demonstrate, beyond question, that these papers were mathematically inconsistent.

What about Experimental Evidence?

Finally, we should discuss the experimental evidence with regard to the existence of tachyons, since physics is, after all, an experimental science. There were no experiments that provided any reason to believe in the existence of tachyons. With no knowledge of their properties (mass, charge, interactions with subluminal matter), it was difficult to design experimental searches. However, there were two somewhat related arguments that raised strong observational doubts concerning their existence. Both grew out of the fact that, unlike an ordinary particle, the energy of a tachyon did not have the same sign in all inertial frames.

First, let's look at the possible radioactive decays of a proton with the emission of a tachyon. We'll work initially in the inertial frame in which the proton is at rest, which we'll call S(rest). Normally, we would say that the decay of such a proton is forbidden by conservation of energy. Initially the momentum equals zero, since the proton is at rest, and the only energy is that associated with the proton mass, $m_p c^2$. The emission of a decay particle will cost the mass and kinetic energy of the decay particle, which are always positive. In addition, since the decay particle will in general have momentum, the proton will have to recoil in the opposite direction so that the total momentum of the two particles together remains zero. This means that the proton will also have nonzero kinetic energy after the decay. Hence, the final energy of the system will necessarily be greater than the initial energy, and the decay will be forbidden by conservation of energy.

The proton itself can't disappear or change its internal state because of another conservation law called the conservation of baryon number. Currently fashionable theories suggest that, in fact, the proton may decay into a positron and other light particles, with most of the mass energy of the proton going into kinetic energy of the decay particles. This process violates the law of conservation of baryon number, since the proton has baryon number and the lighter particles do not. However, since current experiments indicate that the half-life for this process cannot be less than about 10^{33} years, or about 10^{23} times *greater* than the lifetime of the universe, nonconservation of baryon number can safely be ignored for most purposes. This enormous timescale means that the likelihood of an individual proton decaying is extremely small. However, if you look at a large enough number of them, you should see at least a few of them decay. There are ongoing experimental efforts to observe proton decay by looking for an occasional event in very large tanks of water, located in mines deep underground to shield them from so-called background reactions. These are other processes which can look to the experimenters like proton decay. Since a given proton has about a chance of 1 in 10^{33} of decaying in a year, the tank of water should contain at least 10^{33} protons in order to observe roughly 1 decay per year. If you are one of the leaders of an experimental group that is successful in this endeavor, you can safely start packing for a trip to Stockholm and the next Nobel Prize award ceremonies.

However, since the sign of the energy of a tachyon is not Lorentz invariant, if the decay particle is a tachyon it may have negative energy in the proton rest frame, and the earlier argument based on nonconservation of energy does not work. This is because one can always find a negative energy and corresponding momentum for the tachyon, such that the negative energy of the tachyon is balanced by the positive recoil energy of the proton. Therefore, the total energy and momentum are conserved and the emission of a tachyon is allowed. The process could be described by the following equation:

Proton Decay in Rest Frame

$$p(mc^2) \rightarrow p(mc^2 + E_T) + T\,(-E_T)$$

This is an equation of the sort chemists use to describe chemical reactions. Here, p and T stand for proton and tachyon, respectively. The arrow indicates that a process occurs in which the particle or particles on the left side of the arrow are transformed into those on the right. The arrow may be read as "becomes" or "goes to form." Here, mc^2 is the initial energy of the proton and $-E_T$, where E_T is positive, is the energy of the emitted tachyon. Since the tachyon energy is negative, it will be traveling backward in time. Conservation of energy is satisfied, since $mc^2 = (mc^2 + E_T) - E_T$.

If protons at rest decay by the emission of tachyons, the recoiling protons should make tracks in a bubble chamber. These are devices in which moving charged elementary particles leave behind tracks composed of small bubbles. Searches were actually done in stacks of old bubble chamber photographs left over from previous experiments performed for other reasons, looking for tracks left by recoil protons from spontaneous decay of a proton into a proton and a tachyon. (The tachyons might be electrically neutral, in which case they would leave no track in the bubble chamber.) None were found.

Advocates of the reinterpretation principle would say that what would really be seen here was not the emission of a negative energy tachyon, which would travel backward in time, but the absorption by the proton of a positive energy anti-tachyon traveling forward in time. They would say the process observed would be described by the following equation:

Proton Decay in Rest Frame according to Reinterpretation Principle

$$p(mc^2) + T(E_T) \rightarrow p(mc^2 + E_T).$$

In this reaction, an incoming tachyon collides with the proton and is absorbed, transferring its positive energy to the proton. They would argue that nothing was observed, because it was quite possible that there weren't many positive energy tachyons wandering around in empty space. Note the conservation of energy is also satisfied here since the change in sign of the tachyon energy is compensated by the fact that the tachyon energy has been taken from the right (final energy) side of the conservation of energy equation to the left (initial energy) side of the equation. The change in sign and switch in sides of the equation thus compensate one another.

However, there is a problem with this explanation. We can always find a moving inertial frame, let's just call it S'(moving), in which the energy of the tachyon is positive. By the principles of relativity, we know that if the decay is allowed by the conservation laws in S(rest), it will also be allowed in S'(moving). Since this is not its rest frame, the proton will be initially moving and thus will have kinetic energy. In this frame, the decay process will involve the loss of kinetic energy by the proton, with the lost energy being converted into the positive energy tachyon. The reaction would be described by the equation:

Proton Decay in S'(moving)

$$p(E') \rightarrow p(E' - E_T') + T(E_T').$$

The primes indicate that these are the energies in the frame S'(moving) where the proton had a large initial kinetic energy and the tachyon energy E_T' is positive.

Having positive energy, it will be seen to travel forward in time by observers in S'(moving) and thus will unambiguously be seen as an emission process, which does not require the absorption of an incoming particle. The tachyon viewed as emitted in S'(moving) will appear in the proton's rest frame as the required incoming anti-tachyon (don't worry about the "anti" business—we're just being pedantic), whose absorption gives kinetic energy to the proton that was initially at rest in that frame. Thus, it would seem that the existence of tachyons, if they are coupled to protons or other subluminal particles, would give rise to an unobserved decay of those particles into tachyons. In the particle rest frame, this would appear as the absorption of a positive energy particle. One can, of course, avoid any disagreement with experiment by assuming the coupling of tachyons to ordinary matter is sufficiently weak. Of course, if that coupling becomes too weak, the tachyons become essentially unobservable and therefore of no interest.

A similar problem arises in the consideration of high energy cosmic ray protons incident on earth after crossing galactic or intergalactic distances. In this case, the earth plays the role of S'(moving), relative to which decaying particles are in motion. Since the decay of protons at rest into negative energy tachyons is allowed by the conservation laws, at high enough speeds relative to S'(moving), where the decay tachyons have positive energy, the decay will also be allowed. This means that, in the earth frame, the cosmic ray protons will emit positive energy tachyons and lose energy by tachyon emission. The fact that cosmic ray protons are able to retain their extremely high energies over periods of time, probably of the order of millions of years, again implies that if tachyons exist, their coupling to ordinary matter must be exceedingly weak. (You might be wondering why high-energy cosmic rays don't decay into ordinary particles by converting their kinetic energy into the energy of decay products. The answer again lies in the principles of relativity, which assure us that all inertial frames are created equal. Therefore, we can look at the problem in the rest frame of the cosmic ray particle, where it has no kinetic energy, and, as we discussed above, the decay is forbidden by conservation of energy. The point is that if a decay process is forbidden or allowed, respectively, in one inertial frame, the principles of relativity assure us that it is forbidden or allowed in all inertial frames.)

By the early 1980s, the field of tachyon physics appeared to have about run its course. The idea had been interesting and deserved the exploration it received. Most importantly, it led to a wider understanding of the connection between superluminal travel—or, in the case of tachyons, superluminal communication—and the problem of paradoxes associated with backward time travel. However, as far as theory went, tachyons ultimately seemed to leave one with a choice between what were regarded as unacceptable paradoxes or the equally distasteful introduction of a preferred direction in space. Observationally, although this had received less discussion, the existence of tachyons seemed to imply unobserved and unwanted decay processes for subluminal matter. As a result, interest in tachyons declined rapidly and pretty much faded away, fortunately along with the steady stream of tachyon-related manuscripts Allen had been receiving from *Physical Review* to referee.

In fact, the term "tachyon" still appears in the literature in connection with what is called string theory, but in a rather different context. String theory tachyons are quantum states that have a negative mass squared. These states are not, however, taken to be associated with free particles zipping around with superluminal speed. The connection of these tachyons to the tachyonic particles of the kind already discussed is this: if you use the same formula, $E_t = m_t c^2 / \sqrt{1 - \left(v^2 / c^2 \right)}$, for the energy of a tachyon as a conventional particle, then m_t must contain a factor of i and m_t^2 must contain a factor of $i^2 = -1$, because one has the square root of a negative number (thus, a factor of i) in the denominator. We avoided this by taking the denominator to be $\sqrt{(v^2 / c^2) - 1}$ so that m_t can be a real number. The two procedures are in fact equivalent, since the factor of i, when present, always cancels out. Taking m_t to be imaginary does not run afoul of the rule that physical observables must be real, because m_t is not an observable. It is the so-called rest mass, which gives the mass, or equivalently, E_t / c^2 of the particle when it is at rest. But *tachyons are never at rest.*

Allen returned his full attention to elementary particle physics and, in particular, to a newfound interest in the connections between particle theory and cosmology. Happily this forced him to make some efforts to strengthen his rather sketchy knowledge of Einstein's general theory of relativity. This was to prove useful when, about fifteen years later, he found himself thinking again about problems that were familiar from his work on tachyons, but this time largely in the context of general rather than special relativity. Two of his Tufts colleagues were working on these or related questions. One of these, Larry Ford, whose office was next door to Allen's, had begun a very active collaboration with a fellow named Tom Roman.

7

The Arrow of Time

If someone points out to you that your pet
theory of the universe is in disagreement
with Maxwell's equations—then so much
the worse for Maxwell's equations. If it is
found to be contradicted by observation—
well these experimentalists do bungle things
sometimes. But if your theory is found to be
against the second law of thermodynamics
I can give you no hope; there is nothing for
it but to collapse in deepest humiliation.

SIR ARTHUR STANLEY EDDINGTON,
The Nature of the Physical World

Things are as they are because
they were as they were.

THOMAS GOLD

As we discussed in the last chapter, it is a
basic assumption of physics, backed by
very strong experimental evidence, that the laws of physics do not distinguish
between different directions in space. To take a trivial example, suppose we
have a container, which is isolated from the rest of the world and divided in
half by a vertical barrier of some sort, oriented so that the barrier runs north-
south. Suppose we first observe the container at a time $t = -t_0$. At that time,
the western half of the container is filled with air, but the other half has been
pumped out so that it contains vacuum. We also assume that there is a valve in
the barrier that may be opened to allow gas to flow from one side to the other.
If the valve is opened at $t = 0$, then almost at once half the gas will flow from

< 76 >

west to east into the hitherto empty eastern half of the container. Thus, prior to the opening the valve all the gas is in the western half, while in the future of the opening, that is, at $t > 0$, the container will be filled with a uniform density of gas.

What if we repeat the experiment with the initial orientation of the can reversed, so that it is the eastern half which is initially full? Without even thinking about it, we know the answer. The gas will again flow from the full side into the empty side, this time, from east to west, until, after the opening of the valve it is distributed uniformly throughout the container. The laws of physics make no distinction between east and west. However, the gas always goes from a nonuniform distribution in the past to a uniform distribution in the future. We never see a process that would appear to us as a uniform distribution of gas throughout the can turning spontaneously, as time increases, when the valve is opened, into a distribution where all the gas is in just half the can. This is an example of a clear physical distinction between the positive and negative time directions. The laws of physics, thus, do make such a distinction. Physicists and philosophers often refer to such a distinction as an "arrow of time" pointing from the past toward the future whose direction has an origin in the laws of physics.

How does this asymmetry between past and future arise? Surprisingly, the basic equations of physics do not distinguish the negative and positive time directions, that is, the past and future. These equations are Newton's laws of motion for systems that are adequately described by classical mechanics and the corresponding quantum mechanical equations for systems where quantum corrections are important. Both sets of equations possess a property called *time-reversal invariance*. Because of this property, these basic equations do not themselves distinguish between the positive and negative time directions. Newton's laws themselves do not define an arrow of time.[1]

---

1. In the case of quantum mechanics, time-reversal invariance is only approximate. The quantum mechanical equations describing the radioactive decay of certain elementary particles *do* distinguish the two directions of time. These particles have very short half-lives; therefore, they occur only when they are produced at very large terrestrial particle accelerator laboratories or, occasionally, by very high energy cosmic ray particles incident from outer space. When produced, they decay almost at once into more "ordinary" elementary particles that obey time-reversal invariance. It is thus difficult to imagine that the distinction between past and future in the laws governing these objects has anything significant to do with the obvious distinction between past and future which we encounter in our everyday lives (on the other hand, some physicists, such as the mathematical physicist Roger Penrose, believe that nature is providing us here with a very important clue).

Let us examine this property of time-reversal invariance in detail. To begin with, suppose we have a system which contains some number N of particles. We describe the system by giving its initial conditions at $t = -t_0$. The initial conditions are the position and momentum (or equivalently, the velocity) of each particle. This requires specifying a total of 6N numbers, since for each particle we must give its position and momentum in the x, y, and z directions. The laws of physics plus the initial conditions then determine the state of the system for all $t > -t_0$, in particular, at $t = +t_0$.

Now imagine a second system, which we will call the time-reversed system, with the same number of particles. We specify its initial conditions at $t = -t_0$ in the following way. We will take the position of each particle in the new system at $t = -t_0$ to be the same as that of the corresponding particle in the original system at $t = +t_0$. The momentum of each molecule in the time-reversed system, however, is taken to have the same magnitude—but exactly the *opposite* direction—as the momentum of its partner in the original system. In our example, then, the time-reversed system at $t = -t_0$ would be a gas with its molecules distributed uniformly throughout the container, at the same positions as the molecules of the original gas at $t = +t_0$, and with momenta of the same magnitude but in exactly the opposite direction as the momenta of the corresponding molecules in the original gas.

Then the consequence of the property of time-reversal invariance is that, according to Newton's laws, each molecule of the time-reversed system will run backward along the same path followed by the corresponding molecule in the original gas. Watching the actual behavior of the time-reversed system would be indistinguishable from watching a movie or video of the original system being run backward. In particular, at $t = +t_0$, the molecules of the time-reversed system will be at the same position as the molecules of the original gas at $t = -t_0$. Thus, the gas in the time-reversed system will spontaneously flow back until it occupies only half the container!

The fact that the distribution in space of the molecules of a gas in a container becomes more uniform as time increases is thus not a consequence of some property of Newton's laws. Time-reversal invariance shows you how to find initial arrangements of the molecules in a gas which, evolving under Newton's laws, will tend to rush into one region of space, rather than becoming uniform, as time increases.

To illustrate how strange this result is, let's look at another, perhaps more familiar, example. Suppose you have a movie or video of a person diving off a diving board into a pool. Sometimes, as a joke, people run such a thing backward. What you see when you do this is comical, because you see the obviously preposterous sight of the diver emerging feet first from the pool and flying backward to land on the board. Obviously such things don't happen. They never happen. Yet, your authors are claiming that, because of time-reversal invariance, the following is true. Suppose you were to start the physical system of the diver's body plus the water in the pool in the time-reversed state of the actual state after the diver enters the water, which is a conceivable starting state of the system. Then the actual result, according to the laws of physics, would be the same as you see when you run the video of the dive backward. But we've just said that such a thing would obviously never happen. You could certainly be forgiven for thinking that either the authors or the laws of physics have gone off their rocker.

However, the situation is not quite that bad. It is true that, because of time-reversal invariance, the laws of physics do guarantee that there is a state of the gas molecules, that is, a set of initial or starting values for the position and velocity of each gas molecule, which would lead to their spontaneously rushing into half the box. There is even a state of the molecules in the pool and the diver's body that would lead to her finding herself suddenly shot out of the pool and back to the diving board. However, as we're about to see, one of the most important laws of physics tells us that, while in principle these things could happen, as a practical matter they never do and never will. The probability of such things happening is so absurdly small that one would have to wait for a time equal to the age of the universe multiplied by an incomprehensibly large number before ever seeing all the gas molecules in a container of ordinary size spontaneously, as a result of their random motion, rush into half of the container. This law, called the second law of thermodynamics, together with a new physical quantity we have not yet encountered, called entropy, is the subject of the next section. You might, by the way, justifiably ask why we are starting with the *second* law. In fact, you already know the first law of thermodynamics. It's a name often used in the branch of physics called thermodynamics for the law of conservation of energy.

Entropy, the Second Law of Thermodynamics, and the Thermodynamic Arrow of Time

There are two different ways of specifying the state of a system, such as the molecules of a gas in a closed container. What we actually observe about such a system are a few macroscopic (i.e., large-scale) properties of the system. Let's consider gas in a closed container. We'll suppose the gas is in equilibrium, by which we mean that its observed properties are not changing in time. Then the observed state of the system can be specified by just three measurable quantities, which can be taken to be, for example, the volume of the container and both the temperature and the total mass of the gas in the container. The temperature is a measure of the average kinetic energy of the individual molecules as they bounce randomly around inside the container, colliding with one other and the walls as they do so. The total mass of the gas determines the chemical composition of the gas and N, the total number of molecules in the container. (There is also the pressure of the gas, which is the force per unit area the gas molecules exert on the walls of the container as they hit the walls and bounce off. This, however, is not an independent quantity but is determined by a physical law, called the equation of state, from the other three.) The state of the gas, expressed in this way, is called the *macrostate*.

However, while the macrostate specifies what we can easily observe, it is very far from providing a complete description of the state of the gas. To do that, we would have to give the full set of 6N numbers specifying the individual positions and momenta of each molecule. This is called the microstate of the system.

Of course, we can never know the particular microstate of the system, which is, moreover, continually changing in time as the molecules move about and collide randomly. Knowing the macrostate only determines various average properties of the microstate, and there are a huge number of possible microstates that are consistent with a given macrostate. Often, there is nothing in the physics to make any one of the microstates that are compatible with a given macrostate more probable than any other. Hence, each is equally probable, and the total probability of finding the system in a given macrostate is just proportional to the total number of microstates compatible with the macrostate in question; we will call this number n. (Do not confuse N, the number of molecules in the container, with n, the number of possible microstates for some given macrostate. While n depends on and increases with N, the two numbers are in general very different.)

It turns out that for a technical reason it is more useful to deal, not with n, but with a parameter called the entropy, for which we will use the symbol s, which is defined as the logarithm of n, that is, $s = \log n$ (you need to distinguish s from S, which we will continue to use to label an inertial reference frame). One advantage of s is that it is easy to show that, unlike n, the total s for two separate systems is just the sum of the entropy for each one separately. As you probably learned at some point, the meaning of log n is $n = 10^{\log n}$. From the definition you can see that as n gets bigger, so does log n. From the definition, $s = \log n$, the entropy s must also increase as n increases.[2] However, log n is much smaller than n. For example, 1,000,000 is 10^6, so the logarithm of 1,000,000 is only 6. Nevertheless, in cases like the gas molecules in a box, n is such a fantastically large number that the entropy is also very large.

Since n increases when the entropy increases, a macrostate with higher entropy will always be consistent with more microstates, and thus be more probable than one with lower entropy. As time goes on, systems tend to evolve from states of low probability to states of high probability. Therefore, *an isolated system will always go from a state of entropy s_1 at time t_1 to one of entropy $s_2 \geq s_1$, at time t_2, if $t_2 > t_1$.* (The case $s_2 = s_1$ will usually occur only if constraints prevent the system from evolving into a state of higher entropy). That is, entropy increases (or possibly stays the same) as one goes forward in time. This statement is the *second law* of thermodynamics. Note this means that if $t_2 < t_1$, then $s_2 \leq s_1$, since then the entropy cannot decrease in going from t_2 to t_1. Thus, entropy decreases—or possibly stays the same—as you go backward in time.

The second law thus provides an "arrow of time." That is, it distinguishes between the two directions of time. The positive direction in time is the direction of increasing entropy, for example, the direction in which the molecules flow from half the container to fill the whole container. There are *many* more (not just two more, by the way) possible arrangements of the molecules when the whole container is available; that is, the entropy is *much, much* higher, when the molecules occupy the whole container. Similarly, the second law guarantees that the molecules will never flow spontaneously back into just half of the container as time increases, since that would correspond to an entropy decrease. Even though, as we saw, Newton's laws allow microstates of the system that would lead to this behavior, the proportion of such states, and hence, the probability of seeing the system undergo such an event, is so small that it just "ain't gonna happen." If you spend your life waiting around to see an observable violation of the second law, you'll be disappointed.

Due to the increase in entropy mandated by the second law, a system will evolve rapidly until the entropy becomes essentially equal to the maximum entropy allowed by the constraints on the system, such as the size of the container. At that point, further change in the system is inconsistent with the sec-

2. Strictly speaking this is the *natural* logarithm, i.e., the logarithm to the base e, where $e \approx 2.71828\ldots$, but it makes no substantive difference to our argument.

ond law. The state of the system in which the entropy has its maximum value is the equilibrium state, in which the observable properties of the system remain constant. Further observable evolution of the macrostate can occur only if one makes some change in the constraints of the system, for example, by opening a valve. In fact, however, unobservably small violations of the second law do occur continually due to very tiny statistical fluctuations of the entropy away from its maximum value, which quickly vanish as the inescapable hand of the second law makes itself felt.

A system in equilibrium, very near its state of maximum entropy, has no thermodynamic arrow of time. The fact that our world does have such an arrow means that it is very far from equilibrium. Its initial conditions at very large negative time were such that its entropy was very low, and, as mandated by the second law, began to increase with time—an increase that is still, on the average, going on. Thus, we can say that the time asymmetry arises not from the laws of physics themselves but from the initial conditions of our universe.[3]

We will assume for the moment that the gas molecules in our box interact only by direct physical contact, either when they collide with one another or with the container walls. The equilibrium state of the gas is then one in which its properties are uniform throughout. It is relatively easy to demonstrate that uniformity maximizes the number of possible microstates and, hence, the entropy. We've just seen one example of this; let's look at another. Consider a system that initially contains hot coals and cold ice cubes thermally isolated from one another so that heat cannot flow between them. If the insulation is removed, heat flows from hot to cold until the system reaches a uniform temperature throughout. This happens because a state of uniform temperature maximizes the entropy of the combined system. Note that it is the system as a whole that is governed by the second law. The entropy, and thus the number of possible microstates, of the coals actually decreases as they cool, but the number of possible microstates of the system as a whole is increased by removing the constraint that the temperature of the coals be greater than that of the ice cubes.

---

3. Penrose has argued that these initial conditions must have been rather special. He feels that the key question in understanding the ultimate origin of the second law is, *why* was entropy lower in the past? Penrose attributes the second law to conditions on the big bang singularity in which the universe began. For a more in-depth discussion of this issue, see Roger Penrose's *The Emperor's New Mind* (New York: Oxford University Press, 1989), especially chapter 7, and *The Road to Reality* (London: Jonathan Cape, 2004), especially chapter 27. A more recent semipopular treatment is by Sean Carroll, *From Eternity to Here: The Quest for the Ultimate Theory of Time* (New York: Dutton, 2010).

One can, of course, make ice cubes in a refrigerator. During this process the entropy of the refrigerator and the surrounding kitchen is reduced. But this violation of the second law does not occur spontaneously. To bring it about, one must do work to pump heat out of the cold refrigerator into the warm kitchen. To generate the required electric power and deliver it to the refrigerator requires a series of processes, such as burning fuel at the electrical generating plant, which produces more entropy than was lost by cooling the refrigerator. When one does a careful accounting, one always finds that the total entropy of the *entire system* involved increases as time increases.

Before we leave the subject of entropy, we should mention another bit of terminology that is often used. An increase in the entropy of a system is often described as an increase in the system's *disorder*. Another way of saying this is to observe that, as a system's entropy increases, we lose information about the system. When the entropy is low, it means the system is in one of a comparatively small number of microstates, for example, all the gas molecules are known to be in one half of the container. As the entropy increases, we have less and less information about the system. That is, the number of possible microstates of the system becomes larger, and its behavior becomes more random, or more and more disordered. Using this language, one may rephrase the second law as saying that *physical systems become more disordered as time increases*.

To give an example, a rock colliding with a plate glass window causes the glass to shatter into a highly disordered set of glass fragments with many unpredictable details, that is, with many possible microstates. The exact pattern of the glass fragments would be quite different each time any window was broken. The entropy of the glass-rock system thus increases when the glass shatters into disorganized glass fragments. The shattering of the window is thus consistent with the second law. On the other hand, the second law forbids the process in which the glass fragments spontaneously reassemble as the rock leaps from the ground and goes flying off.

It is important to emphasize the inclusion of the word "isolated" in the statement of the second law. This again refers to a system that is left on its own without any interference from the outside world. Otherwise, one can be led into all sorts of misstatements. For example, one (frequently found in pseudoscience books) is that the evolution of life on earth, in which systems with lower entropy (higher complexity) arise from ones with higher entropy (lower complexity), violates the second law! This spurious argument then sometimes is used as a justification for the necessity of a "Creator." The fallacy in this argument, of course, is that the earth is not an isolated system, since it receives energy from an outside source, namely, the sun. The total entropy of the earth increases as it absorbs solar radiation. This is not inconsistent with the *local* decreases of entropy required to produce the evolution of living things of greater and greater complexity.[4]

Cause, Effect, and the Causal Arrow of Time

The arrow of time that we have been discussing, provided by the direction in which entropy increases, is called the "thermodynamic arrow of time." There is a second physical principle, called the "principle of causality," which also distinguishes the two directions in time. The principle of causality states that the laws of physics are such that causes always precede effects in time. When relativity is taken into account, this can be restated more precisely by saying that an event at a given point in spacetime can only have an effect on other events which occur in (or on) the *forward* light cone of that point, as we already discussed in chapter 4. (This assumes there are no tachyons.)

In order to explore the principle of causality more fully and to understand the relation between the two arrows of time, we need to consider the meaning of the terms "cause" and "effect" more carefully. These are words we use constantly and of which we have an intuitive understanding. However, in the context of physics, we need to sharpen that understanding.

Precisely what do we mean by the statement that one event is the cause of another? Suppose some event happens and is then followed by a second event. We may have a tendency to think the first event is the cause of the second. For example, suppose that in the fifth inning of a baseball game, the baseball pitcher for the home team is pitching a no-hit game (that is, the opposing team has not made any hits) and the TV announcer describing the game mentions this fact. If, subsequently, an opposing batter gets a hit, many fans in the audience will be furious, insisting that the pitcher lost his no-hitter because the announcer "jinxed" the pitcher by breaking a time-honored taboo against mentioning a no-hit game in progress.

Are the two events causally related, or is it simply a matter of coincidence? It is impossible to prove. Would the pitcher have gotten his no-hitter if the announcer had kept his mouth shut? We don't know. No-hitters are very rare. Statistically it is likely, if there are several innings remaining, that some batter

---

4. A detailed argument is provided in Penrose, *Emperor's New Mind*, chapter 7, and *Road to Reality*, chapter 27.

is going to get a hit before the end of the game, with or without the intervention of the announcer. These authors (one of whom is an avid baseball fan), at least, are of the opinion that blaming the announcer would be an example of what is called the *post hoc ergo propter hoc* (Latin for "after this, therefore because of this") fallacy. This is the fallacy of assuming that, simply because one event follows another in time, they are causally connected. If you were really determined to establish a causal relationship, you would need to undertake a statistical analysis of the relative likelihood of no-hitters remaining intact in circumstances where the TV announcers have and have not mentioned them.

Suppose that you find identical pairs of events, call them type A events and type B events, always occurring together on a number of occasions, in essentially identical circumstances. Sometimes we say that A and B occur in "constant conjunction." Let's say that every time there's an A, it's followed by a B: the occurrence of an A event is both a necessary and a sufficient condition for guaranteeing the subsequent occurrence of a B event. For example, every time you throw a switch, a light comes on (suppose it's a fluorescent light so that there is a perceptible time interval between throwing the switch and the light coming on, which makes the temporal order is easily observable). We would then begin to feel confident that there was a causal relationship between A and B.

The conjunction between the two events would not need to be absolutely constant (occasionally there might be a power failure or the bulb might burn out). To establish a causal relationship, only a statistically significant correlation between throwing the switch and the light coming on would be required. The definition of "statistically significant" would be somewhat arbitrary. However, in many cases, as a practical matter, there would be no doubt that the correlation was significant. For simplicity we will assume this is the case and continue to speak of "constant" conjunction without worrying about statistical questions.[5]

5. In quantum mechanics, there is a phenomenon known as "entanglement," whereby two components of a quantum system are "linked" in some sense, even over spacelike distances. For example, we can have two particles which interacted at one time, but are now (in principle) arbitrarily far apart. They can have a property called "spin," which according to the laws of quantum mechanics can only have two directions: call them "up" and "down." If the particles are in what's called an "entangled state," then a measurement by one observer on one of the particles is correlated with a similar measurement made on the other particle by a second observer. If one observer measures the spin of particle 1 to be up, he knows the other observer will *always* measure the spin of particle 2 to be down, for example. At first sight, this might look like superluminal signaling, if the particles can be arbitrarily far apart. This, however, is not the case. To use this as a signaling system, say to type out a series of dots and dashes, an observer would have to be able to *control*

Having satisfied ourselves that there is a causal relation between A and B events, we now ask ourselves, "Which is the cause and which is the effect?"[6] We're immediately tempted to answer that question by saying that obviously A is the cause since the A events precede the B events. But by proceeding in this way, we would be reducing the statement that the cause always precedes the effect to a mere matter of definition rather than a fundamental physical principle. If we believe that there is a fundamental principle of causality, according to which the cause always precedes the effect, there must be a way of distinguishing cause from effect in some way other than by their temporal order. We cannot do this if, in fact, A and B always occur together, so that A is both necessary and sufficient for B. Constant conjunction is a symmetrical relationship. If A occurs B occurs, and vice versa. The only distinction between A and B, and hence, the only way of saying which is cause and which is effect, is their temporal order.

To have a meaningful principle of causality, we must break the constant conjunction and find a situation where A or B occurs without the other one. Suppose we find that the occurrence of A is a sufficient but not a necessary condition for B to occur. Thus, B always occurs when A occurs, so it is reasonable to say that A causes B. However B can occur without A, so B does not necessarily cause A. In the case of our switch and fluorescent bulb, this would mean that whenever the original switch is on, electric current flows through the bulb. However, it could be that the bulb is also connected to a second source of current through a different switch, and turning that second switch on causes the bulb to light without the first switch being thrown. In this situation then, we can say that event A, throwing the first switch, causes the effect B, the bulb lights. We have thus identified the cause and effect without any mention of temporal order. We can now ask, as a question of experimental fact, whether the cause, as we have identified it, occurs before or after the effect. One finds, of course, that in situations of this sort the cause always precedes the effect.

---

ahead of time which way his spin measurement will come out, and thus what the other person sees. The laws of quantum mechanics—notably the uncertainty principle—make this impossible, even in principle. So although we can say that the two particles are correlated, we can't really say that the measurement of one *causes* what happens to the other. When the observers get together and compare notes *after* the experiment, they will find that every time the first observer measured particle 1 with spin up, the other observer measured particle 2 to have spin down. Sure is strange, though!

6. The following discussion is largely based on a talk by Roger B. Newton, which is published in *Causality and Physical Theories: Conference Proceedings No. 15 of the American Institute of Physics*, edited by William B. Rolnick, 49–64 (New York: American Institute of Physics, 1974).

The laws of physics thus embody a principle of causality and give rise to a causal arrow of time whose direction is such that the cause always precedes the effect. Another way of saying this would be to say that the causal arrow of time points from an event into its *forward* light cone, and that an effect always occurs in (or on) the forward light cone of its cause.

From what we have said so far, it is possible that the thermodynamic and causal arrows of time could have pointed in different directions. It is an experimental fact that this does not happen.

The Cosmological Arrow of Time

There is still a third[7] arrow of time, which takes its direction from the evolution of the universe as a whole. Observation shows that the distance between any given pair of objects in the universe, for example, a pair of galaxies, is increasing with time. In other words, the universe is expanding. Thus, we can introduce a third, or cosmological, arrow of time, which is defined to point in the direction in time in which the size of the universe is increasing. Experimentally, this is the positive time direction, which is the same direction as the thermodynamic (and causal) arrows.

Is this coincidence, or could we have predicted it? We know that the thermodynamic arrow, because of the second law, points in the direction in which entropy increases. Since the cosmological arrow points in the direction of expansion, one would be tempted to say that it also obviously points in the direction of increasing entropy, because the cosmological expansion is just like the gas molecules whose entropy increases when they expand to fill the whole container. While the conclusion is correct, the reasoning required to reach it is more complex in the cosmological case because the universe is a more complex system.

When we were talking about the container of gas molecules, we made a simplifying assumption. We didn't make much of it at the time or explain why we were making it, and you may well not have noticed it, but we said we would assume that the gas molecules interacted only by direct physical contact with

---

7. There are a number of other of arrows of time we have not discussed. Penrose mentions 7. For a more detailed technical discussion, see his "Singularities and Time-Asymmetry," in *General Relativity: An Einstein Centenary Survey*, edited by S. W. Hawking and W. Israel, 581–638 (Cambridge: Cambridge University Press, 1979). A more popular account of some of the arrows of time can be found in Paul Davies, *About Time: Einstein's Unfinished Revolution* (New York: Simon and Schuster, 1995), chapters 10 and 11.

each other or the walls. In particular, we were excluding the possibility of any long-range force acting between distant molecules. In that case it was obvious that the entropy increased, that is, there were more possible microstates available after the gas expanded, because then there was a wider range of possible position coordinates available to the gas molecules.

The same is true in the case of the expanding universe. But in that case, there is a long-range force between the particles—namely, the force of gravity—that, in contrast with the can of gas molecules, plays a significant role. Because of the opposing force of their mutual gravitational pull, as the particles in the universe expand, they are also slowed down. We must remember that the number of possible microstates, that is, the entropy, depends both on the range of possible positions and also on the range of possible speeds, or equivalently, momenta which the particles can have. The increase in the range of positions and the decrease in the range of possible momenta tend to balance one another in their effect on the entropy of the expanding universe, so it is no longer so obvious that the increasing entropy required by the second law will lead to expansion. More sophisticated theoretical analysis—beyond the scope of this book—is required. As an observational fact, however, we do see the universe is expanding, so that the thermodynamic and cosmological arrows do point in the same direction.[8]

8. For more on the deep questions concerning the relation between the cosmological arrow of time and the second law, see Penrose's *Emperor's New Mind* and *Road to Reality*, and Carroll's *From Eternity to Here*.

8

General Relativity
Curved Space and Warped Time

Now I'm free, free-fallin' . . .

TOM PETTY, "Free Fallin'"

In this chapter, we discuss Einstein's greatest achievement—his general theory of relativity. The idea of "curved spacetime" embodied in the theory is crucial for understanding the scenarios for time machines and warp drives that we will discuss in future chapters. We saw earlier that special relativity singles out a particular class of reference frames for describing the laws of physics. These are the so-called inertial frames. An observer in such a frame cannot tell, from measurements made entirely in her own frame, whether her frame is absolutely at rest or moving uniformly. However, the observer *can* tell whether she is accelerating (with respect to an inertial frame). Einstein wondered why there should be such a dichotomy. Why should *any* frame, inertial or accelerating, have a privileged status for describing the laws of physics? He was also aware of the fact that, while Maxwell's laws of electricity and magnetism are the same in all inertial frames of reference, Newton's law of gravitation is not.

Gravity versus Electromagnetism

This difference can be illustrated in the following way. Suppose you have two electric charges some distance apart. Now you suddenly move one of the charges a certain distance from its original position and stop it. How long does it take for the second charge to know that the other charge has changed position? According to the picture of electromagnetism proposed by Michael

< 89 >

Faraday (still essentially the one used today), the space between the two charges is not empty. Each charge produces an electric "field" around itself and can also respond to external electric (and magnetic) fields. (These are the electric and magnetic fields mentioned on chapter 3 in connection with Maxwell's equations.) The two charges in our example interact by means of their electric fields. The field is the intermediary that transmits an electric force, a push or a pull, from one charge to another. If one charge is suddenly moved to a new position, the field around that charge must "readjust" itself around the new position of the charge.

So another way of asking our question is, how quickly does the second charge "know" about the rearrangement of the field of the first charge? Maxwell's equations, which are rigorous mathematical laws describing the behavior of classical electric and magnetic fields, give an unequivocal answer. When the first charge is suddenly moved and then stopped, a "kink" is produced in its electric field. This kink propagates from one charge to the other at the speed of light in the form of an electromagnetic wave (a wave, as discussed in chapter 3, consisting of oscillating electric and magnetic fields). These waves can propagate because changing electric fields produce magnetic fields and vice versa. Therefore, the amount of time it takes for one charge to know that the other has moved is (roughly) the distance between them divided by the speed of light. To sum up, Maxwell's theory is a "field" theory. Charged particles produce electric and magnetic fields in the space around them and interact with one another via these fields.

Newton's theory of gravitation is not like this. It's what is known as an "action-at-a-distance" theory. If we ask the same question for two masses in Newton's theory, we get a very different answer. According to Newton, the space between the two (assumed uncharged) masses is empty. If we suddenly move one mass, the other mass knows *instantaneously* that the other has moved. As we saw earlier, such instantaneous signaling is incompatible with the special theory of relativity, in which the upper speed limit for any signal is the speed of light. Einstein was profoundly bothered by this fundamental difference in character between electromagnetic and gravitational forces, and so he set out to construct a field theory of gravitation after the manner of Maxwell. Einstein could have chosen to try to resolve the difficulties by trying to adjust Newton's theory to be compatible with special relativity; however, he chose a radically different path.

There is another important difference between electromagnetism and gravity. When an electric or magnetic force acts on an object, the resulting acceleration depends on both the mass and the charge of the object. Objects with different "charge-to-mass ratios" will be accelerated differently (this is the principle behind a device known as the "mass spectrometer"). For example, to test whether there is an electric field in some region of space, we could release a number of "test" particles with different charge to mass ratios and observe their accelerations. The situation with gravity is quite different. All objects are affected by gravity in exactly the same way. More precisely, the gravitational force accelerates all objects in the *same* way, regardless of their mass or composition. This is a remarkable feature of the gravitational force, which distinguishes it from all other known forces. Let us delve into this point in a bit more detail.

Mass and the Principle of Equivalence

There are two properties associated with the idea of the "mass" of an object. One is the inertial mass, which is a measure of how an object responds to a force, that is, a push or a pull. More precisely, the inertial mass is a measure of the resistance of an object to a change in its state of motion, that is, its resistance to being accelerated. A Mack truck has a larger inertial mass than a Volkswagen, which is why if you push the truck and the Volkswagen with the same amount of force, you see more change in the motion of the Volkswagen than the Mack truck. Another property is the "gravitational mass" of an object, which is a measure of the ability of the object to produce and respond to a gravitational force. These two properties are associated with the name "mass" but are quite different from each other. Yet, it turns out that these two kinds of mass are equal to one another: the inertial mass of an object is equal to its gravitational mass. There is no apparent reason why this should be so. (The equivalence of the inertial and gravitational mass of an object has also been experimentally tested to great accuracy, to about 1 part in 10^{-12}.) As a result, when one writes down the expression for the acceleration of an object under the influence of a gravitational force, the mass of the object cancels out from the two sides of the equation, and the resulting expression for acceleration is independent of the mass of the object.

Newton's second law relates the inertial mass m of an object to the acceleration a it experiences due to a net external force F acting on it: $F = ma$. Newton's law of gravitation states that the gravitational force F_g felt by an object with mass m due to another mass M is given by: $F_g = -\dfrac{GmM}{r^2}$, where r is the distance between the masses and G is Newton's gravitational constant. The minus sign in the equation indicates that gravity is always attractive. If the net external force acting on the mass m is F_g, so that $F = F_g$, then we have that $ma = -\dfrac{GmM}{r^2}$. The mass m cancels out from both sides of the equation, so then the acceleration due to gravity produced by the mass M is simply given by $a = -\dfrac{GM}{r^2}$. If, for example, M represents the mass of the earth, then this equation says that the acceleration experienced by an object of mass m, due to the earth's gravity, is independent of m. Hence, all objects undergo the *same* acceleration under gravity. (This was illustrated in Galileo's famous—although probably apocryphal—experiment of dropping two spheres of different mass at the same height from the Leaning Tower of Pisa and showing that they hit the ground at the same time.) Newton considered this fact a mere coincidence, but Einstein argued that the equivalence of inertial and gravitational mass was a deep feature of nature, which he elevated to the "principle of equivalence."

In what he called "the happiest thought of my life," Einstein realized that a man falling off a roof will not feel his own weight during the fall. (Landing is of course another matter!) That led Einstein to conceive another of his famous "thought experiments." Consider a person in a rocket ship out in empty space, far away from any gravitating body and with its engines shut off and no rotation. If the person takes out a pocket watch and releases it, the watch will remain in position— the watch, the person, and everything in the spaceship will "float" relative to the walls of the ship. Now, Einstein said, consider a person in an elevator car near the surface of the earth for which the elevator cable has snapped. Such a person will also "float" relative to the walls of the elevator car. Moreover, if he takes out a pocket watch and lets it go, it will stay there—it will "float" relative to the person and the walls of the car! That's because the watch is falling at the same rate as the person and the elevator car. So, during the fall, the person will feel as though gravity has been "turned off." (This is the reason astronauts in the space shuttle are said to experience "weightlessness." They and the space shuttle are both accelerating toward the center of the earth with the same acceleration. This is the acceleration due to gravity, which is necessary to keep the shuttle moving in a circle rather than going off in a constant direction, that is, along a straight line, as it would do if there were no gravitational force pulling inward toward the earth.) Einstein argued that no experiment done *inside* a closed elevator car could determine whether the car was in space far from all gravitating bodies or freely falling near the earth's surface (this is illustrated in figure 8.1).

Einstein then extended his thought experiment further. Suppose an elevator car out in empty space is accelerated at a rate of 32 feet per second squared

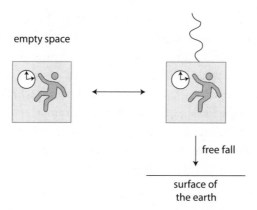

FIG. 8.1. Principle of equivalence I. The behavior of objects inside a freely falling elevator car is indistinguishable from those in an identical elevator car out in space, far away from all gravitating bodies.

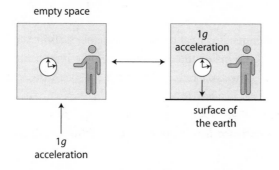

FIG. 8.2. Principle of equivalence II. Objects inside an elevator car accelerating upward at behave the same way as inside a (small) elevator car on the surface of the earth.

(this is 1 *g*, the rate at which objects fall near the surface of Earth). The person in the car takes out his pocket watch and releases it. He sees it fall to the floor of the car at a rate of 32 feet per second squared. As seen from inside the car, the watch and all other objects in the car will behave as though the car were sitting at rest on the surface of Earth (see figure 8.2).

An observer (not shown) standing at rest relative to the earth and watching the situation in the left half of figure 8.2 would see the watch momentarily floating in space and the floor of the car accelerating up to meet it. (Note that the surface of the earth is drawn as flat in these figures for a reason. More about this in a little while.) Einstein's conclusion was that "locally," that is, within the closed elevator car, an observer could not tell whether the elevator car was sitting on the surface of the earth, or out in empty space far away from all gravitating bodies and accelerating at a rate of 1*g*. (Here we assume that the car is "small" and falls for a "suitably short" period of time. We will make these ideas more precise a little later.)

Gravity and Light

Einstein then considered the behavior of a beam of light inside the elevator car in each case. Suppose a tiny horizontally mounted laser attached to the wall of the car emits a light beam. We describe the subsequent motion of the beam *as seen by an observer inside the closed elevator car*. On the left of figure 8.3, the elevator car is unaccelerated in empty space, so the beam simply travels in a horizontal line across the car. On the right, the elevator car is freely falling. Since an observer inside the car cannot tell whether he is in free fall or drifting in space, everything inside the car—including the light beam as well as the observer—falls at the same rate. So again the observer will see the light beam move across the car in a horizontal line.

On the left of figure 8.4 the car is accelerating upward at a rate of 1*g*. Because of the upward acceleration of the car, everything *inside* the car, including

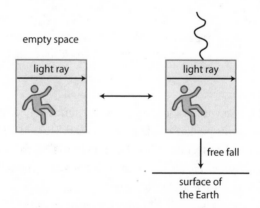

empty space

light ray

light ray

free fall

surface of
the Earth

FIG. 8.3. Principle of equivalence III. The light ray moves in a straight line for both observers.

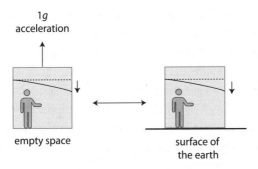

1*g*
acceleration

empty space

surface of
the earth

FIG. 8.4. Principle of equivalence IV. Both observers
must see the light ray bend. Hence, the principle of
equivalence implies that "gravity bends light."

the light beam, appears to observers in the car to accelerate *downward* at a rate
of 1*g*. As a result the beam does not travel in a horizontal line relative to the
car but appears to bend downward, striking the opposite wall at a point below
the point of impact of the initial horizontal path. Einstein drew a remarkable
conclusion from this. If in fact, no experiment done from within the car can
determine whether the car is being accelerated in empty space or sitting at rest
on the surface of the earth, then in the second case (illustrated in figure 8.4)
the path of a beam of light must also bend in the earth's gravitational field.
Gravity "bends" light! One might have guessed this from the fact that light
is a form of energy, and energy has mass ($E = mc^2$ again), and mass is affected
by gravity. Therefore light should have mass and should be affected by gravity
as well. However, one needs to be a bit careful with this argument, as it is not
entirely correct as it stands (which we will discuss later).

Since the gravitational force, unlike the electromagnetic force, affects all
objects equally Einstein reasoned that it might therefore be more appropriate
to describe gravity in terms of the geometry *of* space and time rather than as
a separate force acting in space and time. Furthermore, he argued that all reference frames, both inertial and noninertial, should be on the same footing,
that is, equally valid for describing the laws of physics. He called this idea the
"principle of general covariance."

Tidal Forces

We previously argued that an observer cannot "locally" tell the difference between the effects of gravity and acceleration. Since, unlike electromagnetism,

gravity accelerates all bodies equally, one could not unambiguously deduce the presence of a gravitational field by releasing test masses in a local region of space and time. Put another way, the gravitational force can always be "transformed away" by switching to a (small) freely falling frame of reference where it feels like gravity has been turned off. Note that we have been careful to repeat words like "locally" and "small." Let's see what happens if we remove these restrictions. In special relativity (i.e., in the absence of gravity) we can make an inertial frame of reference as large in both space and time as we like. What if we try to do the same thing near a massive body like the earth? In figures 8.1–8.4, the surface of the earth has been drawn as a horizontal line. That's because we were implicitly assuming that the size of our elevator car was small compared to the radius of the earth. Let's see what happens if we make the car bigger.

First let us point out that an object in free fall near the earth is falling toward the center of the earth's gravitational attraction, which is the center of the earth. If the object is small, compared to the radius of curvature of the earth, and falls for only a short time, then the difference in gravitational force acting on different parts of the object is slight and can be considered negligible. In figure 8.5, we see a very long, horizontally oriented elevator car that is freely falling near the surface of the earth. Two ball bearings start out at each end of the car. Since the car is freely falling, so are the balls, and each ball is falling

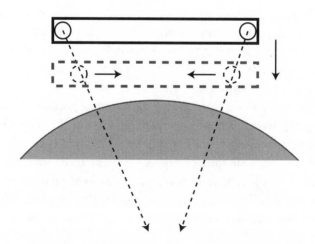

FIG. 8.5. Tidal effects I. An observer inside a long, horizontal, freely falling elevator car will see the two ball bearings move toward one another.

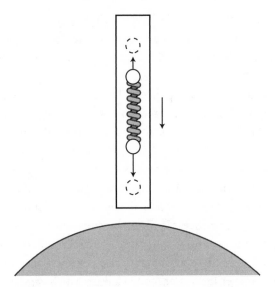

FIG. 8.6. Tidal effects II. An observer inside a long, vertical, freely falling elevator car will see the two ball bearings move away from one another, resulting in a stretching of the spring connecting them.

toward the center of the earth. However, because of the earth's curvature, the paths of the balls are not parallel to one another. Therefore, a part of the balls' motion will be along the horizontal direction. This will have the effect of pushing the balls closer and closer together as the elevator car falls. (Notice that this effect gets smaller as you decrease the horizontal dimension of the car. When the car is very small, compared to the radius of the earth, the paths of the balls are nearly parallel and their horizontal motion is negligible.)

In figure 8.6, we see a long elevator car dropped along its vertical axis. Near the center of the car are two balls connected by a vertical spring. As the car falls, so do the balls, but the lower ball falls at a slightly faster rate than the upper ball. This is because it is slightly closer to the center of the earth than the upper ball and therefore experiences a slightly stronger gravitational force. (According to Newton's law of gravitation, the gravitational force is proportional to one over the square of the distance from the center of gravitational attraction. So if you are twice as far away, the force is $\frac{1}{2^2} = \frac{1}{4}$ as strong, i.e., it decreases by a factor of 4. If you are twice as close, i.e., half as far away, the force changes by a factor of $\dfrac{1}{\left(\dfrac{1}{2^2}\right)} = 4$ and is 4 times stronger.)

Since the upper ball is accelerated at a slightly slower rate than the lower ball, the net effect is that as time progresses the spring connecting the two balls will stretch. If the time of fall is short enough, this stretching will not be noticeable.

Let us put these effects together and consider the free fall of an initially spherical object. We deduce that the difference in gravitational force over different parts of the object will tend to gradually distort the sphere into an ellipsoid as it falls. These differences in gravitational force are called "tidal forces." (They are the same forces responsible for causing the tides in the earth's oceans. The tides arise from the difference in gravitational pull across different parts of the earth exerted by both the moon and the sun.)

The effect is most noticeable for the oceans, since they are the easiest to move around, but the earth's crust "flexes" a bit as well. (On Jupiter's moon Io, this "tidal flexing" is so great that it keeps the interior of Io hot enough to cause the volcanic eruptions of molten sulfur, which were first observed on the *Voyager* flybys.) So the lesson we have learned is that while locally (i.e., in a small-enough region of space and time) the gravitational force can be transformed away by going to a freely falling frame of reference, that frame cannot in general be arbitrarily extended in space or time. Put another way, the gravitational force at a point does not have absolute meaning, but *variations* in gravitational force are detectable. Hence, in the presence of a "true" gravitational field, inertial frames can only be "local inertial," that is, freely falling, frames of reference.

To see this, imagine taking the different possible freely falling frames near the earth and trying to "knit" them together into one single, globally inertial frame. First consider the (artificial) example of a uniform gravitational field. Imagine lots of freely falling elevator cars, representing local inertial frames, located at different points in space. In a uniform gravitational field, all the elevator cars will fall at the same rate, hence, they could be knitted together to form one large elevator car—as large as we like (a "global inertial frame"), which falls at the same rate as the individual cars.

However, real gravitational fields are not uniform. They are only *approximately* uniform over regions of space and time that are small enough for tidal effects to be negligible over the time of the experiment we happen to be doing. But globally they are not uniform, because the gravitational field of a massive object varies in strength and direction at different distances from the object. Now imagine a series of tiny elevator cars distributed around the earth at different distances from the center. Each small elevator car represents a local inertial frame, but since the strength and direction of the gravitational force vary at each point, each car experiences a different gravitational force. Therefore, in this case we cannot knit them together to form one single large ("global") inertial frame that falls at the same rate as the individual frames.

Gravity and Time

In chapter 3 we talked about the problem of synchronizing two clocks at different locations. We more or less took it for granted that once the clocks were synchronized, they would continue to agree, running at the same rate. Of course, that's not necessarily true in practice, but at least in principle, one can imagine making it true by using two identically constructed clocks, or better yet, two atomic clocks making use of radiation from the same kind of atom. While that's true for clocks at rest in an inertial frame, it is not true, as we are about to see, for clocks at rest in a noninertial (i.e., accelerating) reference frame, or, because of the principle of equivalence, for clocks in a gravitational field.

The principle of equivalence can be used to deduce the fact that gravity affects the rate of a clock. Consider two observers, Allen and Tom, stationed at the bottom and top, respectively, of a closed elevator car that is accelerating upward at a constant rate of 1g in empty space.[1] The distance between Allen at the bottom and Tom at the top is h. Let's assume that Allen has a clock programmed to emit light pulses at regular intervals, given by T_{allen}. Tom receives the signals at an interval given on his own identical clock by T_{tom}.

First let's consider the case when the elevator car is moving inertially, that is, at a constant velocity v with respect to an external inertial frame. By the principle of relativity, Allen and Tom could just as well assume that they were at rest, and so the time interval between pulses as measured by Tom's clock would simply be $T_{tom} = T_{allen}$; Tom's clock would receive the pulses at exactly the same rate as Allen's clock emits them.

Let us look at this same situation as seen by an observer in an inertial frame external to the elevator car. The pulses travel at a speed c relative to the inertial observer, as required by the first principle of relativity. This observer will see Tom's clock "running away" from the light pulse at speed v, and hence, the external observer will see the light pulses moving at speed $c - v$ relative to the car.

---

1. We will assume that the velocity of the car during the time of our experiment is always very small, compared to the speed of light, so that we may ignore the effects of special relativity. So for the current argument, Newtonian physics will suffice.

(As discussed previously, relativity only requires that an inertial observer see light moving at speed c relative to *herself*. She may see a light pulse moving at a speed $c-v$ relative to some other object that is also in motion relative to her.) Each of the pulses will thus take a time $h/(c-v)$ to reach Tom at the top of the car. Since v is constant, the travel time for every pulse is the same, and Tom will thus see the interval between the arrival time of Allen's pulses to be the same as that between their emission. Again, we conclude that $T_{tom} = T_{allen}$.

Now let's imagine that the elevator car is accelerating at a constant rate of 1g. Look at the situation from the external inertial observer's point of view. The situation will be the same as before, except that the average value of v during the flight time of a light pulse is now slightly larger for each successive light pulse, and thus the average value of $c-v$ is slightly less because of the acceleration. The top of the elevator is running away from the light pulses at a faster and faster rate, and thus, the travel time for each successive pulse will be greater than that for the one before by an amount which we might call T_{dif}. Suppose, as before, that Allen's clock emits light pulses at intervals of T_{allen}, as measured on his clock. Now the difference in arrival time of successive pulses at Tom's clock will be $T_{allen} + T_{dif} = T_{tom}$. Therefore the time interval between the pulses according to Tom's clock will be *greater* than that measured by Allen's clock, that is, $T_{tom} > T_{allen}$. Hence, Allen's clock is running slow, compared to Tom's clock, since Tom's clock registers a greater interval of elapsed time than Allen's clock.

By the principle of equivalence, the situation we have just described is identical to the case of the same elevator car sitting at rest on the surface of the earth (as far as Allen and Tom are concerned). Therefore, Allen's clock will run slower than Tom's clock in this case, as well. Gravity "slows down" time! Here we assume that the car is small enough that tidal effects are negligible, that is, we assume a uniform gravitational field, but the effect exists for nonuniform gravitational fields as well.

At this point you might be wondering, "Well, if this is true, how come I don't have to reset my watch after visiting the Empire State Building?" The answer, of course, is that the effect is extremely tiny over height differences near the surface of the earth, because earth's gravitational field is extremely weak. "Extremely weak? Oh yeah? Then how come we don't go flying off it?" Well think about it this way: the gravitational pull of the entire Earth on a paper clip can be countered by the pull of a dollar-store magnet (whew, almost dated ourselves and said "dime-store").

Einstein's journey from principle of equivalence thought experiments to the final field equations of general relativity was long and arduous. We will not recount that here, since there are many detailed treatments of the subject, but we will summarize the results of the journey and their implications for us.

General Relativity

Einstein's crown jewel is the general theory of relativity, his theory of gravity. He discovered that what we perceive as gravity can be described as a "curvature" or "warping" of the geometry of spacetime. In the absence of gravity, spacetime is flat and particles and light rays move in straight lines. When gravity is present, particles and light rays move along the closest analogs to straight lines, known as "geodesics." These are the straightest possible lines one can have when spacetime is curved. (For example, on the curved spherical surface of the earth, the geodesics are portions of "great circles." These are circles, such as the equator, which lie in a plane containing the center of the earth. The shortest distance between any two points on the earth's surface is along the great circle joining them.)

A simple two-dimensional example is instructive. Consider a rubber sheet stretched out flat. Roll a marble along the sheet. It moves on a straight line. However, if a bowling ball is placed in the middle of the rubber sheet, the sheet is no longer flat, at least in the region near the bowling ball. A marble rolling along this sheet toward the bowling ball will move along a curved path, following along the "straightest" path it can in a geometry that is no longer flat. The bowling ball determines the geometry of the rubber sheet, which in turn determines the allowed paths of marbles rolling on the sheet. The two-dimensional rubber sheet represents three-dimensional space, with one dimension suppressed. Think of it as a "snapshot" of space at one instant in time. The warping of the sheet in the presence of the bowling ball illustrates the curvature of space in the vicinity of a massive body, such as a star (see figure 8.7). The darker region at the bottom represents the matter of a star, while the light gray region represents the curved empty space outside the star (there is a warping of time as well, but that is not represented in these examples). The third dimension helps us visualize the curvature of the two-dimensional space but it is not part of that space. Similarly, to visualize the curvature of three-dimensional space, we would need a four-dimensional space from which to view it. Since most of us have trouble visualizing things in higher dimensions, the two-dimensional rubber sheet pictures are a useful intuitive crutch, as long as we don't push them too far.

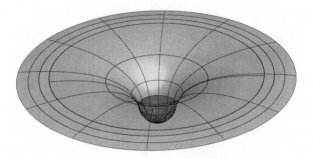

FIG. 8.7. Curved space—a "snapshot picture" of space at one instant of time. The figure depicts the curved space around a massive spherical body, such as a star. Three-dimensional space is represented here as a two-dimensional rubber sheet. Each of the circles on the sheet represents a sphere in three-dimensional space. The space surrounding the sheet has no physical meaning; it simply allows us to visualize the curvature. The warpage of time is not shown in this figure.

In Einstein's general theory of relativity, the presence of matter or energy distorts the geometrical structure of spacetime, much as the bowling ball distorts the rubber sheet. Particles and light rays moving in curved spacetime follow geodesics, the straightest possible paths available to them, just as the marble moves along the straightest path it can on the curved rubber sheet. Newton would say that the earth is held in orbit due to a gravitational "force" of attraction exerted by the sun on the earth. Einstein would say that the mass of the sun curves the spacetime in its vicinity and the planets move along the straightest possible paths in this curved spacetime. The late physicist John Wheeler summarized this by the dictum that "spacetime tells matter how to move; matter tells spacetime how to curve." So gravity is reduced to geometry—a simple and beautiful insight. Einstein's "field" equations are mathematically very complex. However, their physical content can be (very) loosely expressed as

"geometry" = "matter and energy."

At this stage it is important to mention several caveats. Our "cartoon equation" above is highly simplified and does not give the full content of the Einstein equations. First of all, stresses and pressures in the matter, as well as matter flows, also contribute to the right-hand side. Second, only a part of the curvature of spacetime is contained in the left-hand side. There are solutions of Einstein's field equations of gravitation which are "vacuum solutions," representing the warping of empty space. One example is a gravitational wave, which is a ripple of curvature in spacetime that moves through space at the speed of light.[2] Another is the curvature of the region of empty space outside a massive body, such as a star. The first solution of Einstein's equations to be discovered was of just this type. The solution discovered in 1916 by Karl Schwarzschild, aptly called the Schwarzschild solution, describes the curvature of spacetime outside of a spherically symmetric body, again like a star. It covers only the region of empty space outside the star. The spacetime curvature *inside* the star itself will depend on the star's interior structure. Finally, we should note that the field equations themselves are not "derived" any more than are Newton's laws. They are postulates about the way nature behaves—postulates that ultimately must be verified by experiment and observation.

The Three "Classical" Tests of General Relativity

Einstein himself suggested three ways that the theory might be tested. These are now referred to as the three "classical" tests of general relativity, although there have been numerous others since then. One was the successful explanation of an anomaly in Mercury's orbit, called the "precession of the perihelion"—an effect that had been known since the nineteenth century, but that, hitherto, had not been satisfactorily explained. The perihelion is the point in a planet's orbit when it comes nearest the sun. Mercury's orbit is known to be somewhat odd in that its perihelion would slowly shift, or "precess." Most of this shift can be attributed to the gravitational tugs on Mercury from the other planets, notably Jupiter. But when these effects are taken into account, there is still a small amount of perihelion shift that was unexplained by Newton's theory of gravity. Einstein calculated Mercury's orbit in the curved spacetime around the sun, using his field equations of general relativity, and found that the remaining shift was automatically accounted for in his theory.

A second was the prediction of the "gravitational redshift" of light escaping from a massive body. A ray of light "fighting against" the gravitational pull of a massive body loses some energy, which corresponds to a decrease in the frequency of the light. (The frequency of light, or of a more familiar water wave, is the number of wave crests that pass an observer's position every second.)

---

2. Just as electromagnetic waves are produced when charges accelerate, gravitational waves are produced by accelerating masses.

When visible light decreases in frequency it gets redder in color, hence the name gravitational *redshift*. This effect is related to the slowdown of clocks by gravity, which we discussed earlier: a clock that is closer to a massive body ticks more slowly than a clock which is farther away.

The period of a light wave is the distance between two successive wave crests (the "wavelength") divided by the speed of light. If the frequency is the number of wave crests passing per second, then the period is equal to 1/frequency, that is, the time interval between successive wave crest arrivals (e.g., if the frequency is a hundred million vibrations per second, then the period is 1/[100 million] seconds, or ten billionths of a second). We can regard the period of a light wave as the tick of a clock. Atomic clocks are extremely sensitive and can measure time intervals with a precision down to billionths of a second. The ticking rates of two such clocks can be compared to a high degree of accuracy using a technique in atomic physics known as the Mossbauer effect. In the 1960s, R. V. Pound and G. A. Rebka compared the ticking rates of two identical atomic clocks, one on the roof of a building, the other in the basement. According to general relativity, the clock in the basement should tick a few billionths of a second slower than the clock on the roof. Pound and Rebka measured this effect, which agreed quantitatively with Einstein's prediction.

The third of the effects predicted by general relativity is the bending of light by the sun. A ray of light from a distant star that just grazes the edge of the sun will be deflected by a small angle. Normally the image of such a star would be totally obscured by the much-brighter surface of the sun (the photosphere). However, during a total solar eclipse, the moon passes between the earth and the sun and the shadow of the moon covers up the photosphere, albeit very briefly. During this short period of time, stars near the edge of the sun would be visible. Einstein suggested that photographs of stars near the sun be taken during a total solar eclipse. These could then be compared with photographs taken of the same stars when the earth is in another part of its orbit, that is, when an observer on earth can see these stars directly without the sun in the way. An overlap of the photographs should show a shift of the stars' positions.

This is somewhat difficult to do, as the predicted effect is very tiny because the sun's gravitational field, even near its surface, is relatively weak. The amount of shift is an angle of about 1.7″, that is, 1.7 "seconds of arc." One second of arc is about the angular (apparent) size of a tennis ball held at a distance of 8 miles away! Another practical problem is that total solar eclipses tend to be visible in rather inconvenient places, like deserts. During the eclipse, the surrounding air temperature drops, which can cause contraction of telescope equipment, which can also muck up the observation of the effect. There were a number of eclipse expeditions that failed for various reasons. Finally in 1919, the effect was observed by one of the most famous astronomers of the day, Sir Arthur Stanley Eddington, and his measurements agreed with Einstein's prediction. (In actuality, there were fairly large errors in these early experiments, which were later corrected by better equipment and techniques. However, the effect has since been measured numerous times, and the results agree with Einstein's prediction.)

Before he had the full field equations of general relativity, Einstein calculated the bending of light using a principle of equivalence argument similar to what we discussed earlier, in relation to figure 8.4. It's just as well that previous eclipse expeditions were foiled for one reason or another, since Einstein's earlier prediction of the light deflection was off by a factor of 2. It turns out that the other part of the effect comes from the contribution due to the warpage of time in the sun's gravitational field. Had the earlier eclipse expeditions succeeded, it could have been an embarrassment for Einstein. Sometimes it pays to be in the wrong place at the right time! Eddington's announcement of his eclipse results made Einstein a worldwide celebrity overnight, a state of affairs Einstein never understood. Once, Charlie Chaplin invited Einstein to a screening of his movie *City Lights*. The huge crowds that turned out were there to see Einstein as much as Chaplin. Einstein supposedly turned to Chaplin and asked, "What does all this [public adulation] mean?" The more worldly Chaplin replied, "Nothing."

All of these effects we've discussed are tiny within our solar system. However, there are objects in the universe whose gravitational fields are so strong that these effects are enormously magnified. One such object is the remnant that is left when a star like our sun dies. It is called a "white dwarf"—an object with the mass of the sun compressed into a volume the size of the earth, and its density (mass / unit volume) can be hundreds of thousands to millions of times the density of water. A cupful of white dwarf material would outweigh a dozen elephants. Another related object represents the fate of stars whose masses are much greater than that of the sun. Such a star starts off with a much bigger mass, but in its death throes, the core of the star can collapse rapidly under its own weight while the rest of the mass of the star is blown into space. The result is one of the most violent events known in nature—a "supernova" explosion, in which a single star can briefly outshine an entire galaxy of billions of stars. If the final mass of the collapsed core is between two and three times the mass of our sun, the object becomes a neutron star. This is a star made of neutrons, compressed into the size of a city like Manhattan. Its density is comparable to that of matter in an atomic nucleus, so dense that a clump of neutron star material the size of a sugar cube would weigh more than the entire human race! If you could stand on the surface of a neutron star (not recommended!), you could see the back of your head. This is because the bending of light effect is so large near the star's surface that a light ray from the back of your head could be bent in a circle around the neutron star to reach your eye from the opposite direction. Thousands of white dwarfs and neutron stars have been discovered, in our galaxy and in others.

The strangest object of all is the result of the death of a star whose final mass is more than three times the mass of the sun. General relativity tells us that no force can hold the star up against its own gravitational pull, and it must collapse into a "black hole." What this means is that the star collapses to the point where a ray of light emitted from the surface of the star gets immediately dragged back in. The star becomes shrouded by an "event horizon." Just as a ship on earth that passes below the horizon cannot be seen, the region inside of the event horizon is cut off from the outside universe. This is because inside the horizon, an object or light ray would need to travel at a speed greater than c to escape to the exterior universe.

All the effects of general relativity that we discussed earlier are greatly magnified near a black hole. In addition to the bending of light, the gravitational time dilation effect is one of the most dramatic. Again consider the two observers, Allen and Tom. Tom is stationed on the surface of a star which is about to collapse to a black hole—ready to take the ultimate fall for all mankind. Allen, more sensibly, floats in a rocket ship very far away from the star, to watch the action at a safe distance. Tom and Allen synchronize their clocks before the collapse, and Tom agrees to send laser signals to Allen at a rate of once a second, as measured by Tom's clock. The collapse begins. As the star moves inward, Allen begins to notice that each successive signal from Tom takes longer and longer to arrive and is progressively redder in color (gravitational redshift effect). Moreover, Tom's clock is running slower and slower as measured by Allen's clock (gravitational time dilation). By Allen's clock, Tom and the surface of the star take an *infinite* amount of time to reach and fall past the event horizon. By contrast, it takes a *finite* amount of time, as measured by Tom's clock, to cross the event horizon and reach the center. This is warping of time with a vengeance! This scenario might lead one to believe that what Allen will actually *see* visually is Tom and the star moving slowly and yet more slowly, finally freezing at the event horizon after an infinite amount of time. This is *not* the case; it is also a point that, rather annoyingly, most science fiction writers get wrong. Due to the escalating redshift, the light from both the star and Tom's laser will rapidly get shifted out of the range of visible light, to progressively longer and longer wavelengths, and quickly become undetectable. So what Allen will actually see is the star and Tom go dark and "wink out," leaving a central region of blackness, after a very short time. As an example, for a star with a mass of 10 times the sun's mass (10 "solar masses"), Allen would see Tom and the star disappear after only about a thousandth of a second after the onset of collapse. After collapse, the size of the resulting black hole— the size of the event horizon—is given by the "Schwarzschild radius":

$$R_s = \frac{2GM}{c^2},$$

where M is the mass of the collapsed object, G is Newton's gravitational constant, and c is the speed of light. For a 10 solar mass star, R_s is about 20 miles. Anything that falls through the Schwarzschild radius is forever cut off from the outside universe. Numerous objects believed to be stellar mass black holes have been discovered.

Let us now consider the fate of an observer who falls into an already-formed black hole, in light of our earlier discussion of tidal forces; refer to figures 8.5 and 8.6 (for those of you who remember the original *Saturday Night Live!* program, this scenario might be entitled, "Mr. Bill Takes a Trip to the Black Hole"). When the observer is far away from the black hole, he simply feels a rather comfortable free fall. However, as he gets closer to the black hole he begins to feel a stretching force between his feet and his head, and a compression force that squeezes him horizontally. These tidal forces, which we discussed earlier, are the result of the differences in gravitational force between his head and feet, and between the two sides of his body. On a weakly gravitating body like the earth, these differences are miniscule, which is why we don't notice them in everyday life. But near a compressed object, such as a neutron star or a black hole, the differences in gravitational pull even across the size of a human body can become enormous. Finally these forces will be enough to literally tear a human body limb from limb (a process we technical types refer to as "spaghettification").

For a black hole which formed from a stellar collapse, these forces would be strong enough to kill a human being before reaching the horizon. After passing through the horizon, even the (now very deceased) observer's atoms will eventually be torn apart and encounter infinite tidal forces at the center of the black hole. However, the location of this "kill-zone" for a human being depends on the mass of the black hole. The tidal forces are proportional to $\frac{1}{M^3}$, where M is the mass of the black hole. So, somewhat nonintuitively, the tidal forces near and just inside the horizon are smaller for *larger* black holes. An observer falling into a hole of several billion solar masses formed, say, by the collapse of a galaxy of billions of stars, could in principle survive the plunge through the horizon and live for a time inside the black hole. (There is extremely good evidence that such so-called supermassive black holes, with masses of millions or billions of solar masses, lurk in the centers of many, if not all, galaxies, including our own Milky Way.)

Another useful way of representing curved spacetimes is to use light cone diagrams. We can depict flat spacetime by drawing a set of light cones, such that all have their axes parallel to one another and are all the same size (to get the idea, take a piece of paper, roll it up into a cone, and look down the axis of the cone from the top to see the tip at the center). If you imagine looking down from the top of the cones, you would see that all of the cones are circular and that the tips of the cones lie at the centers of the circles. By contrast, one way to represent, for example, the curved spacetime around a black hole, is to draw the light cones in a distorted way compared to their flat spacetime representation.

In figure 8.8, we show the light cones near a black hole. Far from the hole, the light cones on the left look pretty much like their flat spacetime counterparts. As we get closer to the Schwarzschild radius, we see that the cones gradually begin to tip inward. (Here, r represents the radial coordinate.) Right at the Schwarzschild radius, we see that the outgoing leg of the light cone (i.e., the one pointing away from the black hole) is vertical in the diagram. This indicates that a ray of light emitted right at the horizon would take an infinite amount of time to escape. Inside the horizon, for $r < \frac{2GM}{c^2}$, both the ingoing and outgoing parts of the light cone point inward, toward the center. Since any observer's worldline must always lie inside the local light cone, this implies that, once inside the horizon, all observers must inevitably fall toward smaller values of r.

Figure 8.9 illustrates the orientation of the light cones during different phases of the collapse of a star to a black hole. The vertical dotted lines repre-

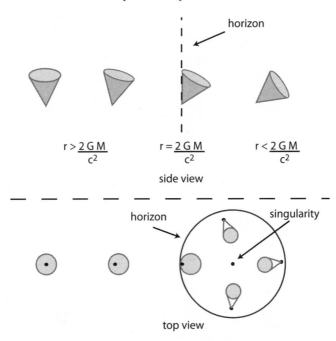

FIG. 8.8. Light cones in the vicinity of a black hole. At the horizon, the outgoing leg of the light cone is "parallel" to the horizon. Inside the black hole, all the light cones point inward toward the singularity.

sent the event horizon; horizontal circles inside these lines represent what are called "trapped surfaces," essentially, regions where light and everything else must unavoidably fall toward the center. The vertical squiggly line at the top of the figure represents the singularity at the center of the black hole, where all matter gets crushed to infinite density and the curvature of spacetime becomes infinite. At the singularity, all known laws of physics break down, to be one day ultimately replaced by the yet unknown laws of "quantum gravity." This would be a theory which merges the laws of the very small (quantum mechanics) with the laws governing the very large (general relativity). We expect that both of these sets of laws must be involved whenever matter is compressed into very tiny volumes and gravitational fields are enormously strong. Although there has been much progress toward a quantum theory of gravity in recent years, it is fair to say that we do not yet have a *definitive* theory.

A black hole can, in principle, be used as a time machine. If we could hover in a rocket ship near a black hole (with our rocket engines turned on so that we don't fall in), our clocks will tick more slowly, relative to the clocks of observ-

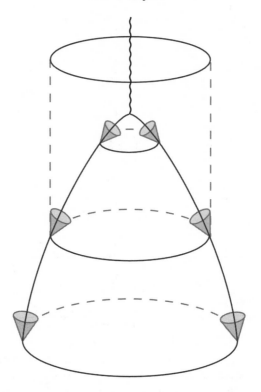

FIG. 8.9. Light cones during the various stages of gravitational collapse.

ers who are far from the black hole. So we could imagine a journey where we travel to a black hole, hover just outside the horizon for some period of time, and return. Since our clocks when we were near the hole ran slow compared to faraway clocks, we will have aged less than our counterparts who did not make the journey. The time difference will depend on our distance to the horizon during the hovering phase and how long we stayed there. As you might suspect, in practice, this scenario is not very feasible. To get an appreciable time difference we would have to hover fairly close to the horizon, and the acceleration required to hold us there would be far more than a human being (or most materials) could endure.

However, one does not necessarily have to accelerate to use the black hole as a forward time machine. Instead, one could go into a circular orbit (i.e., freely falling) around the hole. Unfortunately, there is an innermost stable circular orbit around a black hole that is at a distance of 3 horizon radii, that is, $6GM / c^2$.

(Here, we are assuming a static, noncharged, nonrotating black hole.) The significance of this orbit is that, inside it, a freely falling particle will either spiral into the black hole or be flung back out to large distances. Using the geodesic equations for a material particle orbiting the black hole at this closest stable orbit radius, one finds that the time dilation factor is (only) $\sqrt{2} \approx 1.41$. That is, clocks on the orbiting spacecraft will tick about 1.41 times slower than identical clocks on a distant space station far from the black hole. For each year that passes on the space station, only about 0.7 years would pass on the spaceship, a relatively small but noticeable difference. To achieve a larger time dilation factor, one would have to travel within the critical orbit and undergo large accelerations by using one's rocket engines to avoid falling into the black hole.

9

Wormholes and Warp Bubbles
Beating the Light Barrier
and Possible Time Machines

But why drives on that ship so fast
Withouten or wave or wind?

The air is cut away before,
And closes from behind.

> SAMUEL TAYLOR COLERIDGE,
> *The Rime of the Ancient Mariner*

To look back to antiquity is one thing,
to go back to it is another.

> CHARLES CALEB COLTON

Wormholes

In this chapter, we will examine ways of "cheating" the maximum speed limit imposed by the speed of light when spacetime is curved in unusual ways. One example of curved empty space, discussed in the last chapter, is the spacetime outside of a spherical star. Another such example is a "wormhole." Here is a two-dimensional analog: Take a sheet of paper and cut two identical holes out of it. Now fold the paper over on itself, lining up the two holes one over the other. Separate the two holes above one another slightly and imagine them connected by a smooth tube. Label the holes by A and B. Label a certain point near the outside of hole A by a and a similar point near the outside of the other hole by b. An ant crawling on the paper could get from point a to point b in two

< 112 >

ways. If he is not a very smart ant, he can go the long way around, following the bend of the paper from *a* to *b*. If he is a savvy ant, he can take the shortcut by crawling through hole A, down the tube which connects the two holes, and out the second hole to point *b*. In three dimensions, the two circular holes would appear as two spheres. If you step through one sphere, outside observers will see you shortly emerge from the other sphere. If one traveled from one sphere to the other through normal space, that is, without entering either sphere, the distance could be much, much larger. Each sphere is called a "mouth" of the wormhole. The narrowest part of the "tube" connecting the two mouths is called the "throat." The two-dimensional analog of a wormhole, in terms of a "rubber sheet" diagram, is shown in figure 9.1.

The shortcut through space provided by a wormhole is, effectively, a means of faster-than-light travel. Imagine the wormhole connecting the earth to a star 10 light-years away. A beam of light, traveling in the space outside the wormhole, would require 10 years to make the trip from the earth to the star. (Imagine the upper and lower rubber sheets connected by a strip so that they are both parts of the same sheet.) By contrast, an observer could make the trip in a much shorter time by traveling through the wormhole. This would appear to violate the special relativistic speed limit, which prohibits exceeding the velocity of light. A person going through the wormhole could arrive at the star before a light beam traveling the outside route. However, the person can never arrive

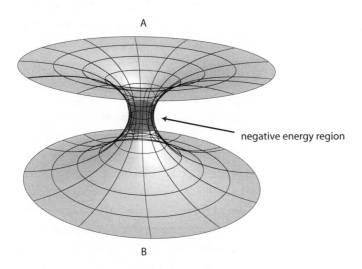

FIG. 9.1. A wormhole. Light rays converge on entering the top mouth and diverge on leaving the bottom mouth.

at the star before a light beam taking the *same route*, that is, through the wormhole. When spacetime is curved, the special relativistic speed limit means that you cannot exceed the speed of light relative to your immediate surroundings.

The idea of wormholes, though in a somewhat different form than we will be talking about, goes back to Einstein and Nathan Rosen in 1935. The concept came up again in the 1960s with the work of Martin Kruskal and that of Robert Fuller and John Wheeler, all at Princeton University. Unfortunately, as Fuller and Wheeler realized, this type of wormhole is unstable—the throat collapses in on itself so rapidly that even a beam of light does not have time to travel through it. Light or anything else falling into such a wormhole gets caught in the "pinch-off" of the throat, where the curvature of space becomes infinite. If that was not bad enough, this kind of wormhole is like a black hole. It has an "event horizon." This means that it would take an infinite amount of time, as measured by outside observers, for you to fall into the wormhole. And once inside, you could never escape. Such a wormhole is "nontraversable"—not a very promising possibility for playing interstellar hopscotch.

For these and other reasons, most physicists did not take wormholes very seriously as objects that might exist in the real world—or that would be very useful even if they did.

As described in the last chapter of Kip Thorne's excellent book *Black Holes and Time Warps* (1994), the situation changed dramatically in the late 1980s when Thorne, who works at Caltech, received a call from his friend, the astronomer Carl Sagan. Sagan was writing his novel *Contact*, later made into a movie with Jodie Foster, and he wanted a believable way for his characters to travel across the galaxy using some kind of spacetime shortcut. In the novel, he initially used a black hole–type wormhole for this purpose, but Thorne pointed out that this would not do, because such a wormhole has the undesirable features we discussed in the previous paragraph. This got Thorne to thinking about exactly what *would* be required to make a "traversable" wormhole, that is, one with no horizons and no pinch-off of the throat, and with properties which would enable human beings to travel comfortably around the universe. Thorne knew that the usual type of wormhole would collapse, but it was a vacuum solution, that is, consisting of only curved empty space with no matter or energy. What if you "threaded" the wormhole with some kind of matter or energy? Would it then be possible to get a wormhole with all the nice properties we discussed?

The way one usually goes about solving Einstein's equations is to assume the presence of a "physically reasonable" distribution of matter or energy, such as a spherical star or a collection of electromagnetic fields or particles, on the right side of Einstein's equations. You then solve (i.e., integrate, for the benefit of those who know some math) Einstein's field equations to find the spacetime geometry produced by that distribution of matter. This is in general a very difficult task, except for cases of high symmetry, such as the case of an exactly spherical object. Thorne and his graduate student, Mike Morris, took the opposite approach, which one might call "geometry first, matter-energy second." They constructed a wormhole geometry that would be suitable for interstellar travel with: no horizons, no infinite curvature, reasonable traversal times, and comfort for human travelers—a "traversable wormhole." Morris and Thorne then put this geometry into the Einstein field equations to find the matter and energy distribution that would give rise to this geometry. This is much easier than the usual approach.

However, there is no guarantee that the matter-energy obtained by this method makes physical sense. If this was all there was to it, Einstein's equations would have no predictive power at all. In fact, one can write down *any* spacetime geometry, "plug it in" to the left side of the Einstein equations, and find the corresponding mass-energy distribution on the right side of the equations, which generates that geometry. Any solution of Einstein's equations corresponds to *some* distribution of matter-energy. With no restrictions, one can get any geometry one likes by assembling the appropriate distribution of matter and energy. We refer to these geometries as "designer spacetimes." However, deciding what constitutes "physically reasonable" matter-energy is not so easy. Einstein's equations by themselves don't tell you this. You have to make some additional assumptions, known as "energy conditions."

The weakest of these assumptions is called, appropriately, the "weak energy condition." Loosely, it says that the density (mass per unit volume) of matter or energy can never be negative, as seen by any observer. Here, "negative" means less than the mass or energy density of empty space. This condition is obeyed by all observed forms of matter and energy in classical physics, that is, when effects due to quantum mechanics are neglected. Energy conditions tell us what are "physically reasonable" distributions of matter-energy. These distributions, in turn, produce what we would then consider to be physically reasonable spacetime geometries. However, the energy conditions themselves are not derivable from general relativity.

Perhaps Einstein was aware of this when he said of his theory:

But it [general relativity] is similar to a building, one wing of which is made of fine marble [left part of the equation], but the other wing of which is built of low grade wood [right side of the equation]. The phenomenological representation of matter is, in fact, only a crude substitute for a representation which would do justice to all known properties of matter.[1]

Morris and Thorne found that the stuff they needed to hold a wormhole open violates the weak energy condition (i.e., it has a negative energy density, at least as seen by some observers), so they dubbed it "exotic matter." This material must have a repulsive gravitational effect on ordinary matter. To see why this must be so, recall the effect that gravity has on light rays. Normal gravitational fields tend to focus light rays, much like a lens. Refer to the wormhole illustration in figure 9.1. Let the heavy black lines in the figure represent light rays falling radially (i.e., toward the center) into one mouth, A, of a wormhole. The rays initially get closer together as they approach the wormhole throat but then diverge (i.e., move farther apart) as they pass through the throat and exit the other mouth, B. This implies that there must be something to counteract the normal tendency of light rays to focus under the influence of gravity. It is the negative energy ("exotic matter") near the throat that provides a repulsive gravitational effect on the light rays, causing them to defocus.

The question of whether the laws of physics allow the existence of exotic matter is a subject we will discuss extensively in a later chapter. For now, let's just assume that we can obtain the exotic matter required and press on to discuss the possible consequences of traversable wormholes.

For a wormhole to be "traversable," in the sense of Morris and Thorne, it has to be comfortable for human travelers. This means that, in addition to having no singularities and no event horizon, the wormhole must have no large tidal forces that could potentially tear a human body to pieces and must have traversal times much smaller than a human lifetime. In their paper, they gave a number of specific examples of traversable wormholes with these properties. One disadvantage of their wormholes is that they were all spherically symmetric with the exotic matter tending to be distributed near the throat. Therefore, an observer traveling through such a wormhole must necessarily pass through the exotic matter that maintains the wormhole against collapse. Since the effects of exotic matter on a human body are unknown, this could be a potential problem. Morris and Thorne suggested that one way to avoid this might be to

1. Albert Einstein, "Physics and Reality" (1936), reprinted in *Ideas and Opinions* (New York: Crown Publishers, 1954), 311.

insert a vacuum tube through the throat that would shield the traveler from the exotic matter.

Matt Visser, now at Victoria University of Wellington in New Zealand, came up with a clever way around this problem. He devised a solution for a *cubical* wormhole. In his wormhole, the exotic matter is confined to the "struts" making up the edges of the cube. As a result, a traveler can enter the wormhole through one of the cube faces without directly encountering any exotic matter. Over the last two decades, Visser has probably contributed more to the subjects of wormholes and time travel than anyone since Kip Thorne. He has written a book for experts on the subject entitled *Lorentzian Wormholes: From Einstein to Hawking* (1995). The book discusses a wide variety of wormhole solutions in addition to the original ones of Morris and Thorne. We shall have more to say about Visser in a later chapter.

Warp Bubbles

In 1994, it was shown by Miguel Alcubierre, then at the University of Cardiff in the United Kingdom, that general relativity also allows the possibility that one could create a "warp drive" with many of the properties of the one seen on *Star Trek*. This consists of a bubble of curved spacetime surrounding a spaceship. In Alcubierre's original model, the ship is propelled by an expansion of spacetime behind the ship and a contraction of spacetime in front. (Later work by José Natário at the Instituto Superior Técnico in Portugal showed that this was not a necessary feature for a warp drive spacetime. In his model, spacetime is contracted toward the front of the ship and expanded in the direction perpendicular to the ship's motion. Natário's bubble "slides" through spacetime by loosely "pushing space aside.") The bubble and its contents could travel through spacetime at a speed faster than light, as seen by observers outside the bubble.

Once again, this might seem like a violation of the ultimate speed limit imposed by special relativity. However, it is important to note that in this case we are dealing with a *curved, dynamic,* spacetime, whereas the spacetime of special relativity is flat and unchanging. The prohibition against reaching or exceeding light speed is in fact obeyed, but not in the way you might expect. Special relativity demands that the spaceship's worldline must always lie inside its local light cone. That is, in fact, true in the Alcubierre spacetime. But because the spacetime is curved in an unusual way, the local light cones inside the warp

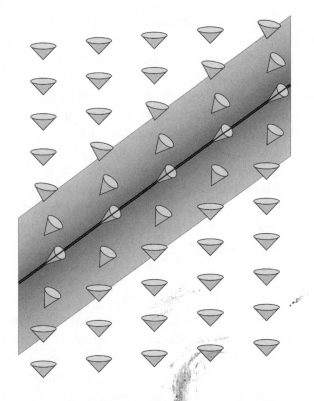

FIG. 9.2. The Alcubierre warp drive spacetime (adapted from fig. 7.2 of James B. Hartle's book *Gravity: An Introduction to Einstein's General Relativity* [San Francisco: Addison Wesley (2003), 145]). Light cones inside the warp bubble are tilted at angles greater than 45° with respect to the light cones outside.

bubble are tilted at an angle with respect to the local light cones outside the bubble. The Alcubierre spacetime is illustrated in figure 9.2. The shaded region represents the "worldtube" of the bubble, that is, its path through spacetime. The thick black line represents the worldline of the spaceship, assumed to sit at the center of the bubble. Note that the light cones outside the bubble are just those of flat spacetime. As we move up from the bottom of the diagram, we see that inside the worldtube of the bubble, the light cones are tilted at angles greater than 45°. However, as you can see, the worldline of the spaceship always lies inside its local light cone, although it is outside the light cone of distant observers. So essentially what we've done here is "speed up" light inside the bubble relative to observers outside the bubble. Observers inside the bubble can thus travel at faster than light speeds relative to observers outside, but still slower than the *local* light speed inside the bubble.

There are a number of other nice features of Alcubierre's model. Spacetime inside the bubble is flat, so the observers inside the bubble are in free fall. They also experience no wrenching tidal forces; all the spacetime curvature is in the bubble walls. In addition, the spacetime is designed in such a way so that clocks inside the bubble tick at the same rate as clocks outside the bubble, so the time dilation problems of special relativity are avoided. Recall in our earlier discussion of the twin paradox, in ordinary flat spacetime, a rocket observer could make a long journey to a distant star in her own lifetime, but when she returns to earth, hundreds of thousands of years may have passed. That's because her clocks and the clocks on earth don't tick at the same rate. The Alcubierre spacetime avoids that problem, so that if you travel from here to a space station near Betelgeuse, your clocks and their clocks are ticking at the same rate and there's no relative aging. Handy, if you want to have a sensible United Federation of Planets (always wondered how they got around that in *Star Trek* . . .). Another advantage is that, in contrast with the case of a wormhole, building a warp drive does not require poking a hole in spacetime, which nobody knows how to do.

One big disadvantage of warp bubbles, as noted by Alcubierre himself, is that they, like wormholes, require the use of "exotic matter." Another distinct disadvantage, first noted by Serguei Krasnikov, then at the Central Astronomical Observatory at Pulkova in St. Petersberg, is that it is not possible for observers inside the warp bubble to steer it! This is because the front edge of the bubble is not causally connected to its interior. To see this subtle but very important point, refer to figure 9.3.

The symbol v in figure 9.3 denotes the speed of the entire bubble, as measured by external observers, while c, as usual, stands for the speed of light in flat spacetime. For superluminal travel, we must have $v > c$. Inside the bubble, the speed of a light beam, v_{beam}, *as measured by observers outside the bubble*, is equal to the speed of light plus the speed of the bubble, that is, $c + v$. That's because all of the contents of the bubble, including the light beam, are carried along at a speed of $v > c$, relative to external observers. Outside the bubble, the speed of a light beam is simply c, relative to these same observers. We expect the beam speed to vary continuously as a function of the distance from the center of the bubble, as we go from the interior to the exterior. Therefore, if the beam speed is $c + v$ inside the bubble, with $v > c$, and drops to c outside the outer bubble

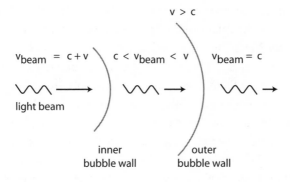

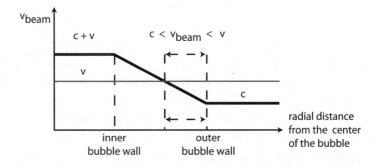

FIG. 9.3. The speed of a light beam in the interior of the bubble, inside the bubble wall, and outside the bubble. The interior of the bubble is causally disconnected from the outer bubble wall.

wall, then somewhere *inside* the bubble it *must* pass through $v_{beam} = v$. This is illustrated in figure 9.3, which plots the speed of a light beam versus the (radial) distance from the center of the bubble. But when the speed of a light beam is $v_{beam} = v$, which it reaches inside the bubble wall, the light beam is traveling at the same speed as the bubble. Therefore, it will simply travel along with the bubble and never make it to the bubble's outer edge. Hence, an observer inside the bubble cannot send a causal signal to the outer wall of the bubble, so that part of the bubble is out of his control. To steer the bubble, a starship captain would have to be able to contact all parts of the bubble. Therefore, Captain Kirk will be in for a surprise when he tells his helmsman, "Hard about, Mr. Sulu, the Klingons are attacking!"

Given the problems with steering the bubble, we might say that riding in an Alcubierre warp bubble is analogous to catching a streetcar. You have no control over the car; you just hope it takes you to where you want to go. Also like streetcars, you can't create and control warp bubbles on demand, they have to be prepared in advance before you can use them, as also pointed out by Krasnikov. Suppose you want to create a warp bubble that will get you from earth to Alpha Centauri 4 light-years away (it's actually about 4.2, but let's use 4 to make the arithmetic simpler) in one day, say, on January 1, 2200. You can't do this by starting on December 31, 2199. By then the spacetime point on Alpha Centauri on January 1, 2200 is far outside your future light cone, and the earliest time at which you can affect anything happening on the star is December 31, 2203. Remember from chapter 4, you can't do anything to affect what happens outside your future light cone. If you want a warp bubble to arrive at the star on January 1, 2200, the latest date at which you arrange for this is Jan. 1, 2196. Starting then, you can in principle arrange for a warp bubble to leave your location, on December 31, 2199, and arrive at the star on January 1, 2200. If you wanted to, you could then arrange for a daily warp bubble service to Alpha Centauri arriving every day *after* January 1, 2200. Of course, all the foregoing is based on one minor assumption: we are supposing that people by 2200 will have learned how to create warp bubbles that can travel faster than light.

A related problem, pointed out by Natário, as well as Chad Clark, Bill Hiscock, and Shane Larson, then all at Montana State University, is that there are horizons that form in front of and behind the starship when the bubble speed reaches the speed of light. The horizon behind the ship consists of a region from which no light rays can reach the ship. The horizon in front consists of a region in which no signal can be received *from* the ship. This is easiest to see if one considers the simple case of the behavior of a light wave traveling directly along the line of the ship's direction of motion. A light wave following directly behind the ship can never catch up to it once the ship reaches and exceeds the speed of light. A light wave emitted from the front of the starship, along its direction of motion, will get outrun by the front part of the bubble, so there is a region in front of the ship which such waves can never reach. Hiscock raised the possibility that these horizons might disrupt the stability of the quantum vacuum around the ship, causing a large "back-reaction" effect on the bubble that would prevent it from ever reaching the speed of light. More recent work by Stefano Finazzi, Stefano Liberati, (from the International School for Advanced Studies in Trieste) and Carlos Barcelo (from the Instituto de Astrofisica de Andalucia in Spain) appears to confirm this idea. (We will have more to say about quantum vacuum back-reaction effects later in connection with wormhole time machines.)

The Krasnikov Tube: The Superluminal Subway

Shortly after Serguei Krasnikov noted the steering problem with Alcubierre's warp drive, he came up with a different model for a warp drive. Rather than using transitory warp bubbles, he suggested one might create tube-like regions of space, reaching, for example, from earth to Alpha Centauri, within which space would be permanently modified to allow superluminal travel in one direction. He suggested that a spaceship crew could journey at sub-light speed from earth to a distant star, modifying the structure of spacetime in a tubular region behind the ship as it went. The modification would consist of "opening out" the *backward* part of the future light cones, that is, the part that points in the direction opposite the direction of the ship's motion. This would be a causal process, unlike trying to steer a warp bubble.

The crew would save no time on the outbound journey, over and above the usual time dilation effects of special relativity, because they would be traveling at sub-light speed during this part of the trip. But on the return journey, depending on how much the future light cones in the backward spatial direction (i.e., along the return path) had been opened out, the ship could travel at arbitrarily high speed and return arbitrarily close to the time it left, as measured by clocks on earth. As a result, the *round-trip* time could be made arbitrarily short!

Unlike the warp bubble where space is modified only temporarily, during the bubble's passage, once the Krasnikov tube has been created, the space within remains modified, and superluminal travel in the return direction of the originating rocket ship remains permanently possible. If riding a warp bubble is like catching a streetcar, then travel in the Krasnikov spacetime is analogous to catching a subway train. As a result, we dubbed it the "superluminal subway," or, the "Krasnikov tube" in our first collaboration, an article in *Physical Review* (1997) elaborating on several aspects of Krasnikov's original work.

The Krasnikov spacetime is illustrated in figure 9.4. The light cones near the edges of the figure are just those of flat spacetime. The thick gray line inclined at a less-than 45° angle represents the worldline of the ship—and the "digging" of the Krasnikov tube—on the outbound journey. The two thinner dark gray lines represent the worldlines of the ends of the tube. The region bounded by the three gray lines is the spacetime history of the interior of the tube (which is why the central white region extends upward in the diagram). The backward parts of the future light cones in this region are opened outward, with a maximum allowed opening of 180°. (Note that the *forward*-pointing

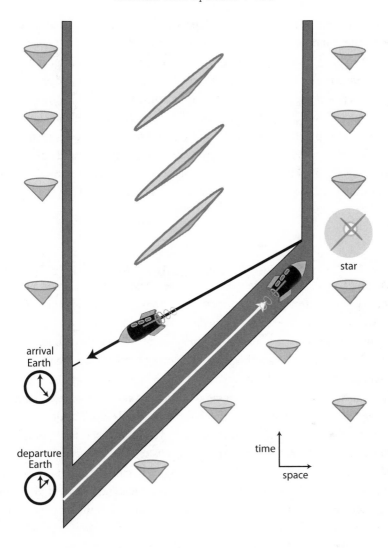

FIG. 9.4. The Krasnikov tube, a superluminal subway. Light cones inside the tube are stretched out in the "backward" direction. As a result, the roundtrip can be made arbitrarily short!

parts of the light cones remain unmodified inside the tube. They are parallel to the forward-pointing parts of the light cones outside the tube. In this latter limiting case, a ship that immediately turns around upon the completion of its outbound journey could travel back along an antiparallel line to its outbound path and reach its departure point arbitrarily close to the time it left. To see this in the figure, imagine the worldline arrow representing the return trip to be parallel but opposite in direction to the thick inclined line. If you then make these two lines arbitrarily close together, corresponding to an arbitrarily quick turnaround time, you could make the total time interval, measured on the earth clocks between departure and arrival, arbitrarily small.

However, as you might suspect by this time, Krasnikov tubes, like wormholes and warp bubbles, also require exotic matter for their construction and maintenance. Additional work by Ken Olum and by Sijie Gao and Bob Wald at the University of Chicago suggests that *any* sort of superluminal travel requires "exotic matter," not just the ones we have discussed.

Wormholes, Warp Drives, and Time Machines

Once you have created one wormhole, warp bubble, or Krasnikov tube, presumably, you should be able to make another. And with two such objects one can make a time machine, in principle. The basic idea is similar to the two-tachyon transmitter-receiver system discussed in chapter 6.

Figure 9.5 shows the space and time axes of two inertial frames that are moving relative to one another (the light gray 45° line represents the path of a light ray, for reference). The events C and D lie on an $x = const$ line and so are simultaneous in the unprimed frame. The events A and B lie on an $x' = const$ line and so are simultaneous in the primed frame, which is moving with respect to the unprimed frame. The events A and B, and C and D, respectively, could represent the mouths of two wormholes. The dashed line paths could represent the paths through the wormholes from A to B, and from C to D. If the internal length can be made arbitrarily short, then A and B could be essentially glued together internally, though possibly widely separated in the external world, and similarly for C and D.

Consider the following scenario. Suppose we succeed in constructing a wormhole that connects events C and D, which are simultaneous in the unprimed coordinate system. The time and space coordinates of C and D in that system will thus be (T,x_C) and (T,x_D), respectively. By the principles of relativity, we can construct a similar wormhole that connects events A and B, which are simultaneous in the primed frame, where they will have coordinates (T',x'_A) and (T',x'_B). Since, according to the Lorentz transformations, the time of an event in one inertial frame depends on both its time *and* position in a different frame, the coordinates of the events A and B in the *unprimed* frame will be (T_A,x_A) and (T_B,x_B), with $T_A \neq T_B$, as shown in figure 9.5.

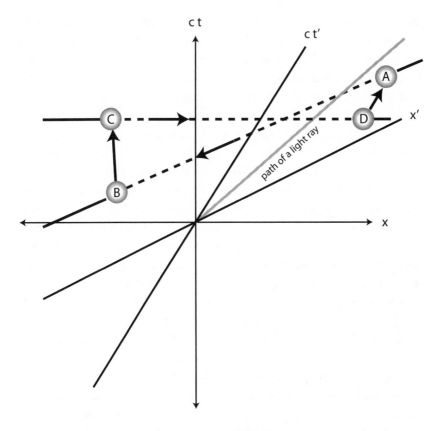

FIG. 9.5. Superluminal travel can lead to backward time travel.

An observer entering a wormhole mouth at A would instantaneously (as measured by her own clock) emerge at B, at an earlier time in the unprimed frame. If she then journeys on a timelike path through normal space from B to C and enters the second wormhole at the mouth located at C, she will find herself instantly emerging from the mouth at D. Note that D is in the past of her departure event A. If she then travels from D to A along a timelike path through normal space, she will arrive at the moment she left. She might then prevent herself from setting out in the first place.

What makes this scenario possible is the relative motion between the two wormholes and the fact that for spacelike paths, unlike for timelike or lightlike paths, the time order of events is not invariant. As a result, even though A and B are simultaneous in the primed frame, B lies in the past of A in the unprimed frame. (Incidentally, the crossing of the two dashed lines is simply the result of restricting our spacetime diagram to one space dimension. To see this, draw the unprimed coordinate system and the points C and D on one piece of paper and the primed coordinate system with points A and B on another sheet parallel to the first. By putting the sheets close together, we can still make the paths BC and DA timelike without having the two dashed lines overlap.)

You may have noticed this scenario is very similar to one in chapter 6 where we showed that tachyons could be used to send a message into the past. In fact, the scenario we have described for creating a time machine would work for other methods of superluminal travel, as well. It depends crucially on being able to causally connect points in spacetime that would otherwise be separated by a spacelike interval. The dashed line between C and D in figure 9.5 could represent the path of a warp bubble, with the dashed line between A and B representing the path of a second warp bubble moving with respect to the first, with both of them moving along spacelike trajectories. By jumping from one warp bubble to the other, one can make a round-trip (this was first shown in a paper by Allen in 1996).

Similarly, the dotted paths could represent two Krasnikov tubes that are moving with respect to one another, with A and B being the ends of one tube and C and D representing the ends of the other tube. One of the tubes would have its light cones opened out in one direction; the second would have its light cones opened out in the opposite direction. By traveling through one tube and then the other, you would always be traveling in a faster-than-light direction (this was shown in a paper written by both authors). Note also that these scenarios require *two-way* superluminal travel in order to be able to make a round-trip, to close your path in both space and time.

Kip Thorne, together with Mike Morris and another of Thorne's students, Ulvi Yurtsever, discovered a second, very ingenious way of making a wormhole time machine using just a single wormhole. Place one mouth, A, of the wormhole on earth. Put the other mouth, B, in a rocket ship and send it off at a speed near the speed of light and then bring it and the rocket ship back to earth. This scenario makes use of the famous "twin paradox" of special relativity, discussed in chapter 5. By jumping into mouth B, one can emerge out of mouth A in the past. Let's see how this works.

In figure 9.6, the gray vertical line represents mouth A of the wormhole, which remains on earth. The curved gray line depicts the worldline of mouth B, which is accelerated to high speed and then eventually returned to earth. Notice that this part of the picture looks just like our twin paradox spacetime diagram (figure 5.3). So clocks just outside the wormhole mouths experience the

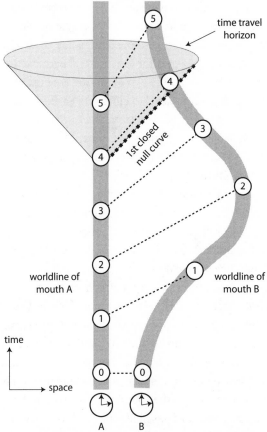

FIG. 9.6. The twin paradox–type time machine of Morris, Thorne, and Yurtsever. The time travel horizon separates the regions of spacetime where time travel is possible from the regions where it is not.

usual time dilation of special relativity. The length of the wormhole (i.e., the distance as measured through the wormhole) is assumed to always be arbitrarily short. (This is not obvious from our discussion. For more detail on this point, refer to figure 14.6 of Kip Thorne's book Black Holes and Time Warps.[2])

The circled points with the same numbers, connected by dotted lines, are at the same proper times, as measured by clocks right at the wormhole mouths.

2. Kip Thorne, Black Holes and Time Warps (New York: W. W. Norton and Co., 1994), p. 501.

So, for example, if you were in the rocket ship at the event labeled "1" at mouth B looking through the wormhole, and if the clock just outside mouth B in the cabin of the rocket ship is reading 1:00, so too is the clock at mouth A at the corresponding event on its worldline labeled "1." Because the length of the wormhole is arbitrarily short, these two points are essentially the same point. Clocks on the spaceship, compared to clocks on earth *as seen through the exterior cabin window*, are time dilated. However, clocks right at each mouth, when viewed *through the wormhole*, read the same time. This means that if you step through wormhole mouth B at event 1, you would emerge at the corresponding event 1 on the worldline of mouth A.

Notice that the dashed paths that connect the similarly numbered points are initially spacelike (i.e., inclined at angles greater than 45°) in the external spacetime outside the wormhole. Therefore, if you jumped into mouth B at event 1 and emerged from mouth A at the corresponding event 1, you would have to travel along a spacelike path *in the external space* to get back to event 1 at mouth B. As we move upward in the spacetime diagram, the dotted paths become progressively less spacelike until we come to the thick dashed line. This critical line represents the first possible closed lightlike (null) curve. A light wave entering mouth B can travel through the wormhole and emerge from mouth A, then travel along a lightlike path in the external space outside the wormhole (along the thick dashed line) and return to its starting point in both space and time. This null curve represents the boundary of the region in spacetime where time travel to the past becomes possible. The light cone of which it is a part is the "time travel horizon," also called the "Cauchy horizon" or the "chronology horizon."

An observer could jump into mouth B at event 4, emerge from mouth A at the corresponding event 4, travel through the external space along a timelike path, and return to her departure point in both space and time. Hence, the events labeled "4" are connected by a *closed timelike curve*. The time travel horizon separates the region of spacetime with no closed timelike curves from the region of spacetime with closed timelike curves. In the region above the horizon, an observer who jumps through mouth B will emerge from mouth A in the past. An observer who initially jumps through mouth A will emerge from mouth B in the future. This type of time travel cannot be accomplished in the usual twin paradox scenario, because, in that case, the two worldlines are not connected through a wormhole.

It is important to note that because of the time travel horizon, *a time traveler cannot return to events that occur before the time machine is first activated*, that is, events that occur prior to the formation of the first closed null curve. Therefore, if the first time machine is first activated in the year 2050, then would-be time travelers cannot return to any time prior to 2050. So you can't use such a machine to go back and hunt dinosaurs (unless some very advanced and much more ancient civilization has built one of these things and left it conveniently nearby for us to use).

Shortly after the article by Morris, Thorne, Yurtsever appeared, other physicists proposed alternative scenarios for turning wormholes into time machines. Igor Novikov, now at the Niels Bohr Institute in Copenhagen and a longtime friend of Thorne, suggested that, rather than moving one mouth of the wormhole away from earth and back again to achieve the necessary time dilation, one could instead whirl it in a circle around the other mouth. Valery Frolov, of the University of Alberta, and Novikov then suggested yet another method: using *gravitational* time dilation to create the time shift between the mouths. One mouth could be placed near a source of high gravity, such as a neutron star, while the other mouth could be placed farther away. Frolov and Novikov showed that, if one waited long enough, such a wormhole would naturally evolve into a time machine.

The effect is the same as in the twin paradox scenario. The only difference is that gravitational time dilation is used instead of special relativistic time dilation to produce the time shift. Their result suggested that wormholes are rather naturally disposed to turning into time machines. Any difference in the gravitational field between the wormhole mouths would have the effect of gradually time dilating one mouth relative to the other, resulting in eventual time machine formation. How rapidly this takes place depends on the size of the difference in gravitational field between the mouths. A wormhole with one mouth placed near the surface of a neutron star would evolve into a time machine much faster than one whose mouth was placed near the surface of the earth.

"Curiouser and Curiouser . . . ": Paradoxes

Once the scenarios for creating wormhole time machines had been proposed, Thorne and his collaborators began to analyze the paradoxes associated with backward time travel. To simplify the problem, they wanted to avoid complicated issues like human free will, which is tricky enough even without the presence of a time machine. So instead of using human travelers, they used billiard balls. Human beings are very complex systems whose behavior is notoriously hard to predict. But the behavior of billiard balls is easily predicted by the laws of classical physics. Thorne and his colleagues studied the billiard ball equivalent of the grandfather paradox.

In the standard paradox, a time traveler goes back in time to shoot his grandfather. As a result of the deed, one of the time traveler's parents is never conceived and so the time traveler is never born. But if he was never born, he could never have built the time machine, gone back in time, and shot his grandfather. So we have the logically inconsistent situation that an event—the grandfather's murder—happens if and only if it doesn't happen. In the usual paradox, we have the possibility that the time traveler may change his mind. For the billiard ball time traveler, there is no mind to change, so the outcome should be unambiguous. A situation like this in which the occurrence of some event, call it event 2 (the time traveler going back in time in the example), causes a another second event, say event 1 (the grandfather's murder), which in turn causes event 2 not to occur, is sometimes called an "inconsistent causal loop." Such loops result in logical paradoxes, and the laws of physics must be such that they do not occur.

The possibility of having such a closed causal loop only arises when backward time travel is possible, that is, when one has a time machine. Conventional physics includes a "principle of causality," according to which effects always follow causes in time. Thus, if event 1 causes event 2, then 2 occurs at a later time than 1 and, thus, 2 cannot cause (or prevent) 1. This principle can clearly no longer be universally true if one has a time machine, since pressing a button on a time machine in the year 2500 might cause a time traveler to appear in, for example, the year 2499.

The billiard ball version of the grandfather paradox is illustrated in Figures 9.7. In this case, the two mouths of the wormhole connect two places at different times. The times shown in the figure are those on clocks outside the wormhole. The terms "older" and "younger" refer to time *as measured by a clock on the billiard ball*. Thorne and colleagues considered the following kind of scenario. A billiard ball is headed for mouth B of a wormhole time machine at external time 3:50 p.m., as shown in figure 9.7a. It enters the wormhole at 5:00 p.m., according to the external clocks. At 4:00 p.m., according to the external clocks, a slightly older (in terms of the billiard ball's own time) version of the billiard ball emerges from mouth A (figure 9.7b), and later collides with the younger version of itself at 4:30 p.m., external time, as shown in figure 9.7c. The collision knocks the younger billiard ball off course, deflecting it so that it does not fall into the wormhole (figure 9.7d). So we have the following situation:

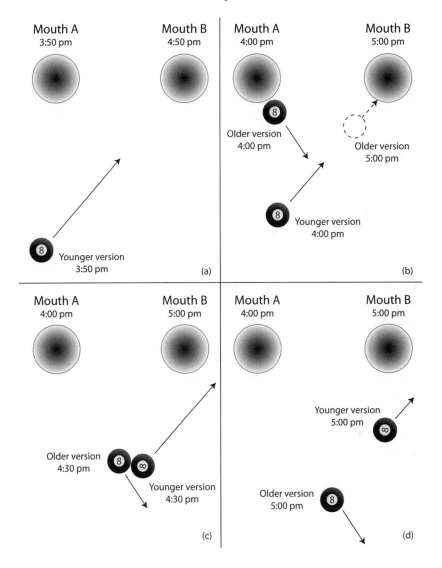

FIG. 9.7A–D. The wormhole version of the grandfather paradox.

A billiard ball falls into a wormhole and emerges an hour into its own past. It then collides with the earlier version of itself, preventing the younger version from entering the wormhole. But if the younger version never entered the wormhole, there would have been no later version to emerge and collide with its younger self. Therefore, the younger billiard ball *would* have entered the wormhole, as there was nothing to deflect it. But if the younger version enters the wormhole, the older version emerges at an earlier time and prevents it from doing so. Therefore, the billiard ball enters the wormhole if and only if it does not enter the wormhole. So we have an inconsistent causal loop, the billiard ball equivalent of the grandfather paradox. In this context, it is sometimes called a "self-inconsistent solution" of the physical equations, such as Newton's laws, governing the motion of the billiard balls.

Let us now consider the situation from the viewpoint of the billiard ball. Imagine there is also a clock attached to the billiard ball, which agrees with the external clocks up to the time the ball enters the wormhole. (We make the realistic assumption that the ball's speed is much less than the speed of light; thus, the effect of the slowing down of a moving clock, predicted by the special theory of relativity, will be completely negligible.) First of all, the reading of the billiard ball clock will continue to *increase* as it goes through the wormhole. If we replace the ball with a time traveler, then what we might call the personal time of the traveler will continue to run *forward* as she travels backward in time *relative* to the rest of the universe. This is, in fact, what is meant by traveling backward in time. She will remember her entrance into the wormhole as being in her own personal past as she emerges from mouth A at an earlier external time (one could also repeat our earlier series of figures, labeling the times on the ball in addition to or instead of the external times).

Moreover, the internal distance between mouths A and B, that is, the distance going through the wormhole, will be much less than the external distance. (That's the whole point of a wormhole.) Therefore, the *elapsed* time on the billiard ball clock will be much less than one hour. In fact, for simplicity, we often picture the internal distance through a wormhole as being essentially zero, although this is not necessary. Making this approximation, going through a wormhole would be like going through a door between two different rooms. However, the clocks in the two rooms would disagree with one another. If we make this assumption that the internal distance through the wormhole is negligible, then one would see the billiard ball clock reading 5:00 p.m., as it emerged from mouth A. If the wormhole is a little longer, so that it takes, let us say, one minute of its own time for the ball to get through the wormhole, the ball clock would read 5:01p.m. as it emerges.

Quantities such as the starting position and velocity of an object are called "initial conditions" by physicists. In the case of the billiard ball scenario, Thorne and his collaborators also found that for every set of initial conditions, such as the starting position and velocity of the ball, it seemed that one could always find a self-consistent solution with no paradox. This is illustrated in

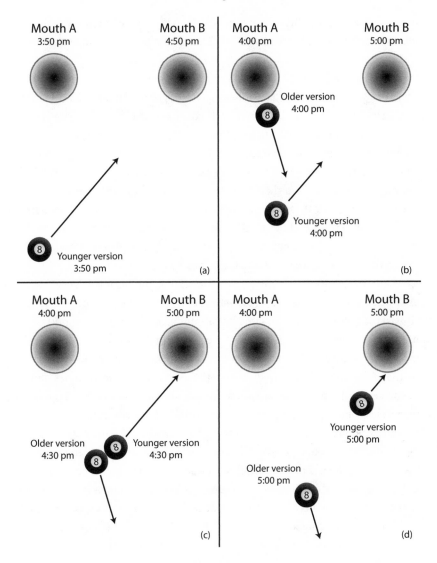

FIG. 9.8A–D. A self-consistent solution to the paradox.

figure 9.8. For example, in the scenario depicted above, suppose the older version of the billiard ball strikes its younger self a *glancing* blow instead of a direct-line collision. If the blow is at just the correct angle, it can deflect its earlier self enough so that it still enters the wormhole mouth B —but at a slightly different point on the wormhole mouth than in our previous scenario. It then emerges at a different angle from mouth A, which causes it to follow just the path required for it to hit its younger self the proper glancing blow to deflect it into just the right point on mouth B. There is no paradox, and this scenario is self-consistent.

So we see that, for the billiard ball model, although there is an inconsistent scenario, given the *same initial position and velocity* of the billiard ball (the same initial conditions), there also exists a self-consistent scenario. This suggests at least one possible resolution to the grandfather paradox. Perhaps in situations where, for the same initial conditions, both inconsistent and self-consistent solutions exist, nature will always choose the self-consistent one. Igor Novikov has championed the idea that the laws of physics allow only self-consistent (i.e., nonparadoxical) solutions. Given a supposed paradoxical scenario, Novikov conjectures that there will *always* be at least one self-consistent solution for the same initial conditions. Novikov and his colleagues examined a variety of models where this indeed seems to be the case.

There was another surprise in store for Thorne and his team. They discovered that there were in fact not just one—but an *infinite* number of—such self-consistent solutions with the same initial conditions! These correspond to the number of times the billiard ball circulates through the wormhole prior to exiting mouth A and colliding with its younger self. This situation of having more than one possible solution is something that occurs in classical physics only in the presence of time machines, when closed causal loops can occur. In general in classical physics, if one knows the position and velocity of all the particles in some system at some time (i.e., the initial conditions), as well as the forces acting on the particle, Newton's laws of motion allow you to determine the subsequent behavior of the particles uniquely (an analogous statement also holds true when the effects of quantum mechanics are taken into account). In the cases where there are a number of self-consistent solutions, Novikov's "self-consistency conjecture" does not by itself tell us which of these is selected by the laws of physics. It only tells us that when confronted with a situation where there are both inconsistent solutions and one or more consistent solutions, the actually observed, physical solution will be a consistent one.

Before leaving this discussion, there is one more subject we should examine. That is the question of conservation of energy. Let us return to figure 9.8. Notice that in figure 9.8a, observers will see only one billiard ball present at 3:50 p.m., while at 4:00 p.m. they will see two (figure 9.8b). The second is the older version of the original ball that has traveled backward in time through the wormhole and reemerged. Since the billiard ball has an energy $m_b c^2$ (neglecting the negligibly small kinetic energy of motion of the billiard ball, whose speed is much less than c), where m_b is the mass of the billiard ball, the appearance of the second ball would seem to indicate a gross violation of the law of conservation of energy. A second ball will be present, and thus, violation of energy conservation will persist until 5:00 p.m. At that time, as seen in figure 9.8d, the younger version of the ball will disappear into the wormhole, and after that there will once more be only a single ball present, as was true before 4:00 p.m.

All our experience indicates that violation of energy conservation is experimentally unacceptable. To avoid it, we must assume that during the external time between 4:00 p.m. and 5:00 p.m., the extra energy represented by the second ball must be compensated by a corresponding *decrease* in the mass of the wormhole due to its interaction with the billiard ball that is passing backward in time through it from 5:00 p.m. to 4:00 p.m. Therefore, if the original mass of the wormhole was M, its mass will be reduced to $M - m_b$ during that period. This will leave the total mass of the wormhole plus two billiard-ball system between 4:00 p.m. and 5:00 p.m. as $2m_b + (M - m_b) = M + m_b$, which was the original mass of the wormhole plus billiard ball. After 5:00 p.m., only the older version of the billiard ball that went through the wormhole is present outside the wormhole. Therefore, the wormhole mass has been restored to its original value of M, so again the total mass of the system has its original, energy conserving, value of $M + m_b$. Thus, the wormhole time machine respects the law of conservation of energy. This is in contrast to H. G. Wells's time machine, which suddenly disappears at one time and reappears at another with no compensating energy increase or decrease in its surroundings.

10

Banana Peels and Parallel Worlds

A paradox, a paradox, a most ingenious paradox.

W. S. GILBERT, *The Pirates of Penzance*

Only one accomplishment is beyond both
the power and mercy of the Gods. They cannot
make the past as though it had never been.

AESCHYLUS

Types of Paradoxes

In this chapter we will discuss time travel paradoxes and their possible resolutions. There are two kinds of paradox that we shall discuss: those we call "consistency paradoxes" and also "information," or, "bootstrap" paradoxes. An example of a consistency paradox is the grandfather paradox (discussed in chapter 9). In this kind of paradox, an event (e.g., the murder of one's grandfather) both happens and doesn't happen, which is logically inconsistent. An information paradox occurs when information (or even objects) can exist without an origin, apparently popping out of nowhere. Let us first consider information paradoxes.

Information Paradoxes

An example of an information paradox is one that we will call "the mathematician's proof paradox." A time traveler in 2040 goes to the library and copies a proof out of a math textbook of a very famous theorem. Suppose that the time traveler then goes back in the past to visit the mathematician who proved the theorem, traveling back to a time *before* he discovered the proof. The time traveler tells the mathematician, "You are going to be famous," and shows

< 136 >

him the proof. The mathematician dutifully copies it down, and subsequently publishes it, establishing his fame in the process. (Of course, the time traveler need not bodily travel back into the past; he could merely send the proof back.) The question is, where did the proof come from? Note that, unlike in the grandfather paradox, everything in this scenario is consistent. The mathematician got the proof from the time traveler, who in turn got it from a book in the library. The real question is: where did the *information* contained in the proof come from originally? Although consistent, this "free lunch" example seems to go against our deeply held beliefs about how the world works. We are not used to information just appearing out of nowhere. (Too bad, it would be a great way to write a PhD thesis!)

Here's another one for you. Carol and Ralph meet one another in the year 2040. Carol says to Ralph, "Go back in time to 2020 and tell my past self that you want to have this meeting with me in 2040. Tell her that these are the instructions she should give you when she meets you in 2040." Ralph uses a time machine to travel to the year 2020 and meets the past version of Carol. He tells her: "Your future self told me that you should meet with me in 2040 and give me the following instructions: 'Go back in time to 2020 and tell my past self that you want to have this meeting with me in 2040. Tell her that these are the instructions she should give you when she meets you in 2040.'" Carol gets into another time machine (or simply waits through the natural aging process) and travels to 2040 where she meets Ralph. Question: Who arranged the meeting in 2040?

"Jinnee Balls" and Clever Spacecraft

In a 1992 article, Lossev and Novikov considered the possibility of self-existing objects, which they called "jinnee[1] balls," which might be associated with time machines. For example, suppose that we have a wormhole time machine. A billiard ball may suddenly exit one mouth of the wormhole, travel through normal space to the other mouth, and enter it, emerging from the first mouth in the past, and so on. All that such an object does is endlessly loop though the time machine. The ball's history has no beginning and no end—it is "trapped" in the time machine.

However, as Lossev and Novikov point out, such an object is forbidden by

---

1. "Jinnee" (also spelled "jinni," "jinn," or "djinn")—is the name of a spirit or class of spirits featured in many Arabian tales. The more-familiar Western spelling is "genie."

the second law of thermodynamics. In order for the scenario we described to be self-consistent, the ball that emerges from the first mouth must be identical in every way to the ball that enters the other mouth. The ball, like all macroscopic objects, will have some temperature above absolute zero, and therefore will radiate heat as it travels. Therefore in traveling through normal space from one mouth to the other, the ball "ages" in the sense that it loses energy in the form of heat. Hence, the ball that emerges from the wormhole mouth in the past cannot be the same ball that entered the other mouth, because it has lost energy during the trip. So, this scenario is not self-consistent.

An example of this kind of inconsistency in science fiction is found in the (rather bad) movie *Somewhere in Time*. A mysterious older woman gives a young playwright an antique pocket watch after seeing the opening of his new play. Years later, he stays in a hotel where he sees an old photograph of that same woman, as she appeared in much younger days. Determined to meet her, he "wills" himself back into the past (huh?), encounters her and (naturally) falls in love. At some point, he ends up giving her a watch (of course, this is the same one she will give him years in the future), and later is involuntarily (and rather inexplicably) snapped back to the future. The watch is a "jinnee" object. But do you see the problem (apart from the mixing of tenses)? Suppose she gives him the watch when it is already an antique, and he keeps it for ten years until he makes his time travel journey. When he gives her the watch in the past, the watch is ten years older (in terms of its own time) than it was when he received it. Suppose she then keeps it for another forty years before giving it to him in the future, when she is an elderly woman. When she gives him the watch, it is then fifty years older (again, in terms of its own time), than when she gave it to him originally. Hence, we have a contradiction. The problem stems from the fact that the watch ages, according to its own time. After the watch makes a time loop, it therefore cannot be the same watch they started with. However, it must be the same watch, if the loop is to be self-consistent.

The jinnee ball scenario could be made self-consistent if the jinnee ball interacts with other objects outside the wormhole, gaining energy from them in such a way as to recreate its initial internal state (i.e., its state when it exited the mouth in the past). This might happen if the ball collides with other balls or interacts with some external energy source. Lossev and Novikov call this a "Jinnee of the first kind,"[2] where matter travels along a time loop. They suggested that the complexity of the "jinnee" that emerges from the time machine may

---

2. A. Lossev and I. Novikov, "The Jinn of the Time Machine: Non-Trivial Self-Consistent Solutions," *Classical and Quantum Gravity* 9 (1992): 2315.

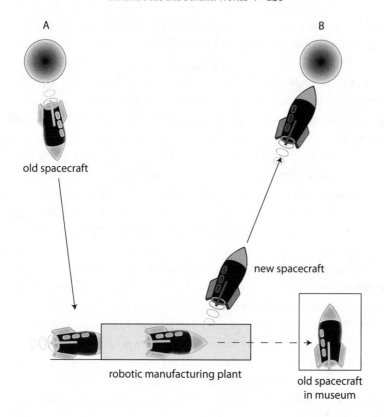

FIG. 10.1. Lossev and Novikov's "clever spacecraft."

be determined by the amount of external energy available to it. The more complicated the object, the more energy is required to recreate its initial state. If we place a large source of energy outside a wormhole time machine, we might see all sorts of complex objects emerge. (In principle, this includes people! Shades of I *Dream of Jeannie!*) However, for a truly self-consistent scenario, the internal state of the ball would need to be reproduced in every detail, that is, the *microstate* and not just the macrostate of the ball (e.g., the temperature). Perhaps the greater the complexity of the jinnee, the lower is the probability of its appearance. If so, then the most likely jinnee objects would be elementary particles.

In the same paper, Lossev and Novikov presented a very inventive example of an "information jinnee" (that they call a "Jinnee of the second kind"[3]), which is

3. Ibid., 2316.

illustrated in figure 10.1. Suppose we know that there is a wormhole time machine somewhere in the galaxy, but we don't know its exact location. We build an automated spaceship manufacturing plant on earth and provide it with necessary raw materials. Then we turn the plant on, withdraw, and simply wait, letting events take their course. (This allows us to avoid the question of free will in this scenario.) From one region of the sky, a very old spacecraft appears from one of the wormhole mouths (mouth A in the figure) and eventually lands on a prepared platform attached to the manufacturing plant. Once there, it dumps its computer core memory into the manufacturing plant's computer.

The memory core contains the specifications for building the spacecraft, as well as the record of its journey, including the locations of the two mouths of the wormhole time machine. From this information, a new spacecraft is built and programmed with the information from the old spacecraft's memory core. The new spacecraft is then automatically launched toward the other mouth of the wormhole, mouth B, whose location was contained in the previous memory core. The old spacecraft is subsequently put on display in a museum.

Note that all we have done is simply to set up an automatic manufacturing plant and provided it with raw materials. What we get in return is the location of a wormhole time machine, the design of a spacecraft, and a very old spacecraft! Lossev and Novikov emphasize that it is *information* that makes the time loop. The old spacecraft ends its life in a museum, so it does not travel along a time loop.

Of course, we don't know if any wormholes exist. Assuming that there is at least one such wormhole in our galaxy, we have no way of reliably calculating the probability per year of such an information loop occurring. This probability might be very small or even zero, or it could be very large. We feel that it should be very small, since we are uncomfortable with the idea of information appearing from nowhere, but we have no entirely satisfactory way of proving this. The fact that a scenario is *consistent* is not a guarantee that it will actually occur. It is equally consistent that nothing happens. When more than one consistent solution is possible, it is not clear how to calculate their relative probabilities.

Wormhole Time Machines and Consistency Paradoxes Revisited

In another 1992 paper, Novikov considered several situations in addition to those discussed in chapter 9, which involved potential consistency paradoxes. As in the case discussed earlier, discovered by the group at Caltech, he showed that consistent solutions could be found. For example, let us suppose that the billiard ball in the example from chapter 9 contains a bomb that will go off and destroy the ball if it is hit by another object. One will then have an inconsistency if the ball goes through the wormhole and comes back and strikes itself even a glancing blow, since the ball will then blow itself to pieces before it can enter the wormhole. Thus, our previous consistent solution has been eliminated. But as Novikov points out, there is still a possible consistent solution. Imagine the ball blows up, producing a rain of fragments that go off in all directions. Some fragments will enter the wormhole at mouth B in figure 9.8 and will emerge at an earlier time from mouth A. It is easy to show that some of the emerging fragments will have the right velocity to hit the incident ball at the moment it explodes and cause the explosion. The explosion is thus its own cause so that we have a self-consistent causal loop and again no paradox arises.

We know that it is possible, then, to find special cases of consistent backward time travel, but we want to know if *all* backward time travel can be made consistent. We will see that, if one has a time machine, it is possible to set up situations where a paradox is inevitable. In these situations it appears impossible, contrary to Novikov's conjecture, to find a self-consistent solution and avoid the paradox. If any such situations do exist, then either we must find a way to deal with paradoxes, or we must conclude that the laws of physics do not allow the construction of a time machine.[4]

One situation where a paradox is inevitable was presented in a paper by Allen titled "Time Travel Paradoxes, Path Integrals, and the Many Worlds Interpretation of Quantum Mechanics," published in *Physical Review* in 2004. Let us go back to figure 9.8. A billiard ball enters mouth B of the wormhole at 5:00 p.m. and reemerges from mouth A an hour earlier. But now let us add a gate through which the billiard ball must pass at 4:30 p.m. in order to get to mouth B (the gate is not shown in the figure). We assume that the gate is initially open so that the billiard ball can pass through, enter mouth B, and emerge from mouth A at 4:00 p.m. However, we will put a detector, say a photoelectric cell, at mouth A that detects emerging billiard balls. When the ball emerges, the detector emits a radio signal to a receiver at the gate, which then causes the gate to close. The radio signal, traveling at the speed of light, arrives at the gate before the ball, so the billiard ball finds the gate closed and

---

4. An interesting example of a (not self-consistent) time travel paradox is presented in the short story "As Never Was," by P. Schuyler Miller (in *Adventures in Time and Space*, edited by Raymond J. Healy and J. Francis McComas [New York: Del Rey, 1980]).

thus never reaches mouth B of the wormhole. But if it never reached mouth B, it could never have emerged from mouth A. Therefore, we have an inconsistent causal loop in which the billiard ball emerges from the wormhole if—and only if—it does not emerge.

In this scenario, the gate is either open or closed. There is not a range of possibilities as in the collision of the billiard ball with itself, where the collision can range from head on to barely glancing. These possibilities allowed Kip Thorne and his colleagues to find a self-consistent solution in the case of the billiard ball that emerged from the wormhole and collided with itself.

However, one has to be careful to be sure that there are really no consistent solutions. In fact, as was pointed out to Allen by the anonymous referee who reviewed the paper on its initial submission for publication, there is a range of possibilities that allows one to find a consistent solution. Suppose the ball arrives at the gate just as the gate is closing and manages to squeeze through but is slowed down by an amount that can be anything, depending on just when the ball arrives in the small time interval in which the gate is closing. It is then possible for the ball to be slowed down just enough so that it arrives at mouth B at 5:30 p.m. rather than 5:00 p.m. It will then emerge from mouth A and be detected one hour earlier, at 4:30 p.m., causing a signal to be sent to close the gate just in time for the incident ball to arrive at the gate and be slowed down as it squeezes through.

However, it turned out to be possible to eliminate the loophole allowing consistency by tweaking the initial setup, as is done in the published version of the paper. In the following discussion, the terms "younger" and "older" refer to time as measured by a clock riding on the billiard ball. Refer to figure 10.2. Let us turn the billiard ball detector at mouth A off at 4:20 p.m., 10 minutes before the younger ball reaches the gate (figure 10.2a). Now suppose the older ball emerges from mouth A at 4:30 p.m. The gate is not closed, since the detector was turned off 10 minutes prior to the ball's emergence from mouth A (figure 10.2b).

Therefore the younger ball would have passed through the gate and arrived at mouth B at 5:00 p.m. It would have then traveled one hour into the past, emerging from mouth A at 4:00 p.m. instead of 4:30 p.m., while the billiard ball detector was still on, thus causing the gate to close (figure 10.2c). The younger ball will find the gate closed and never reach the wormhole, in which case the gate would not have been closed in the first place. Therefore, after the tweaking, one does have a situation where there is an inconsistent causal loop.

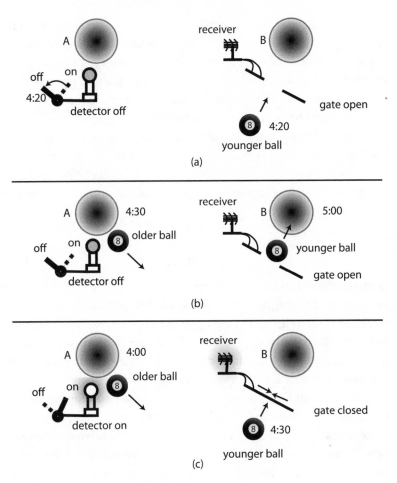

FIG. 10.2A–C. Everett's example of a billiard ball collision scenario with no self-consistent solution.

Our paradox has been restored, and we have thus found a system for which Novikov's self-consistency hypothesis does not hold.

So it appears that, once one has built a time machine, one can find perfectly sensible setups in which a grandfather paradox is unavoidable. One might think it is pointless to even consider the possibility of backward time travel. Conservation of energy says that no process in which the total amount of energy in the universe increases (or decreases) can ever occur. As far as we know, no future civilization, no matter how advanced, will ever be able to create energy. Perhaps the same thing is true of traveling backward in time. Stephen Hawking, among others, believes this idea, which he refers to as the "chronology protection conjecture" (discussed in chapter 12), to be true. On the other hand, as discussed in chapter 9, what we know of general relativity and quantum mechanics seems to at least leave open the door to the possibility of creating wormholes and, thus, time machines.

If backward time travel is possible, there are two general approaches that could allow one to avoid paradoxes. Each of these is illustrated in numerous works of science fiction, but one or the other, at least, must turn out to have a basis in the actual laws of physics, if it turns out that those laws allow one to build a time machine.

Banana Peels

The first of these possibilities is that the laws of physics are such that whenever you go to pull the trigger to kill your grandfather something always happens to prevent it—you slip on a banana peel, for example (we like to call this the "banana peel mechanism"). You *can't* kill your grandfather because you *didn't*. Events that have already occurred cannot be undone. Something will prevent your doing it in your future, after you step out of the time machine, because something did, in fact, prevent your doing it in what is, for the rest of the world, history. A time traveler can be a part of history when he visits the past, but he cannot change that history. He will necessarily do what he has already done, no matter how he tries to avoid it. An excellent fictional illustration of this scheme is Robert Heinlein's classic story *By His Bootstraps*, which is, in our opinion, one of the best time travel stories ever written. A time traveler goes forward in time through a wormhole, decides it was a mistake, and returns to his room through the wormhole in the opposite direction to prevent his initial entry. A struggle ensues in which he inadvertently knocks himself into the wormhole. Other nice examples from television and movie science fiction that we recommend are the *Twilight Zone* episode "No Time Like the Past," and Terry Gilliam's film *12 Monkeys*.

The banana peel mechanism amounts to a slightly revised version of Novikov's consistency hypothesis. It allows the existence of systems that would be perfectly reasonable in the absence of a time machine, and that would lead to a paradox if a time machine were present. But the banana peel mechanism guarantees that if you try to construct such a system with a time machine included, the system you wind up with will necessarily include some sort of unexpected "banana peel" that will avoid the paradox. How might this apply to our system of a billiard ball with a gate, a wormhole, a billiard ball detector, and a signal transmitter? We might find, for example, that when the ball emerges from mouth A, the transmitter that is supposed to send a message and close the gate, thus creating the paradox, develops an unexpected glitch. (Or perhaps the ball may slip on an unexpected banana peel and never enter the wormhole.)

The banana peel mechanism leads to a theory that is, logically, perfectly consistent. It is somewhat (or very?) unappealing, however, because it's hard to understand how the laws of physics can always ensure the presence of a suitable banana peel. There is a troubling fact related to this approach. It turns out that, if one wants to preserve the experimentally verified laws of quantum mechanics, the construction of a time machine in the remote future can affect the probabilities of things happening in the present. For example, the fact that someone is going to build a time machine next week may mean that the probability of your being able to build a properly functioning radio transmitter today (or perhaps of eating your lunch without dropping a banana peel) is, unexpectedly, very low. On the other hand, if no one builds the time machine, the probability that your radio transmitter will function properly and you will throw your banana peel in the garbage can is very high.

Parallel Worlds

The second general approach to paradox-free backward time travel makes use of the idea of parallel worlds. According to this idea, there are two different "parallel" worlds, one in which you are born and enter the time machine, and the other in which you emerge from the time machine and kill your grandfather. There is no logical contradiction in the fact that you simultaneously kill and do not kill your grandfather, because the two mutually exclusive events happen in different worlds that have no knowledge of one another. Like the banana peel mechanism, the idea of parallel worlds is also illustrated in many science fiction works on the theme of time travel. A good example, though one unfortunately out of print at the moment, is the excellent novel *Branch Point* by Mona Clee. Another is *The Time Ships* by Stephen Baxter. This is a sequel to Wells's *The Time Machine*, and the writing style is deliberately—and quite convincingly—a copy of Wells's own.

You could hardly be criticized for saying, "I see that such a theory is logically consistent, but surely the idea of parallel worlds is so outlandish that it should be confined to the world of science fiction." Surprisingly, however, there is an intellectually respectable idea in physics called the "many worlds interpretation of quantum mechanics," first introduced in a paper in *Reviews of Modern Physics* back in 1957 by Hugh Everett. According to Hugh Everett, there are not only two parallel worlds but infinitely many of them that, moreover, multiply continuously like rabbits.[5]

To understand how this works, we need to talk a little about quantum mechanics. It is a theory that can predict only the probabilities of the various possible outcomes of an experiment. It never tells you with certainty what will happen. The probabilities are obtained from what is called the wave function of the object, and the equations of quantum mechanics (with which we will not have to concern ourselves) determine how an object's wave function evolves in time in different physical situations.

Everything we know tells us that quantum mechanics is the physical theory that governs the behavior of all systems, large or small. In the case of macroscopic, (i.e., ordinary, everyday-sized objects) quantum mechanics tells us that the objects behave with essentially complete certainty, as they are predicted to do by classical (Newtonian) mechanics. Hence, we can usually forget the complications of quantum theory in dealing with everyday objects and simply use Newton's laws of motion, which we know empirically work very well for such objects. However, when we deal with atomic- or subatomic-sized objects, we must use quantum mechanics if we wish to get predictions that agree with experimental observation.

Let's consider, for example, an electron. In addition to having a position and a velocity, it may also be thought of as rotating, or spinning, about some axis, like a curveball thrown by a pitcher. According to the rules of quantum mechanics, the electron's speed of rotation can have only one possible value, unlike the baseball's. (The speed of rotation, like many other observable quantities is said to be "quantized," i.e., to have only certain possible values. This is why the theory is called "quantum mechanics.") The only two possibilities for the spinning electron are that its spin may be clockwise or counterclockwise. Suppose that when we first see the electron, it is in a state where its wave function tells us that a measurement of its spin direction will yield clockwise or counterclockwise with probability 2 / 3 or 1 / 3, respectively. Let us now put the electron through what is called a Stern-Gerlach apparatus, which measures the

---

5. The July 2007 issue of the highly respected journal *Nature* contains several articles, many reasonably nontechnical, discussing current views of Hugh Everett's work.

spin direction. Picture the apparatus as having a gauge on it with a needle that initially points to 0. After the measurement, the needle points to 1 if the spin is clockwise and to 2 if it is counterclockwise. Suppose we proceed to make such a measurement and we see the needle point to 1 as it would 2/3 of the time in this situation. In the conventional, or "Copenhagen," interpretation of quantum mechanics, immediately after the measurement the wave function will have changed to one that describes an electron that has probability 1 of clockwise spin and probability 0 of counterclockwise spin. We will find ourselves looking at a gauge whose needle points to 1 in a universe containing an electron spinning clockwise.

In the Copenhagen approach, the microscopic object being measured (the electron) is treated as a quantum mechanical system described by a wave function. However, the large measuring apparatus is regarded as a classical system whose behavior can be adequately described by classical Newtonian physics. In practice this works very well, and there is no difficulty in distinguishing which part of the system we are looking at; it is the measuring apparatus that we will treat classically. As a matter of principle, however, there is no really satisfying way of making this separation. Physicists tend to be satisfied as long as we have a theory that works, in the sense of allowing us to make physical predictions that agree with experiment. We tend to leave such matters of principle to philosophers of science to worry about. In his 1957 paper, however, Hugh Everett argued that, in a really correct version of quantum mechanics, the experimental apparatus should be treated quantum mechanically in the same way as the object being studied. To do this, he developed what is now called "the many worlds interpretation" of quantum mechanics.

In the many worlds interpretation, when a measurement is made, the following occurs: After measuring the electron spin, the measuring apparatus, including the observers looking at it, are in two different states. With a probability of 2/3, you will find yourself in a state (or "world") with a gauge whose needle points at 1 and an electron spinning clockwise. But there will be a second "world" with observers looking at a gauge whose needle points at 2 and where the electron is spinning counterclockwise, and you will have one chance in three of ending up in that world.

More generally, in the many worlds interpretation, whenever a measurement is made, the universe branches so that there is a separate world for each of the possible outcomes of the experiment (often there will be many more than two) allowed by the rules of quantum mechanics. In each world the measuring apparatus will indicate one of the possible outcomes of the experiment, and the measured quantity will have the corresponding value. There will be a copy of the observer in each world who will be looking at the gauge and seeing it have the reading corresponding to that particular world. Our colleague Larry Ford likes to say that the good news about the many worlds interpretation is that you always win the lottery. The bad news is that the probability of winding up in a particular world is equal to the probability, as calculated from quantum mechanics, of obtaining the corresponding result when the measurement is made. Hence, the probability of being the lucky "you" who winds up in the world where you win the lottery is just the usual probability of winning the lottery. So don't pack your bags for that 'round-the-world trip just yet.

Notice that what we are talking about is called the many worlds *interpretation* of quantum mechanics, not the many worlds *theory*. That is, in the absence of a time machine, the Copenhagen and many worlds interpretations—at least in the view of a majority of physicists—lead to identical experimental predictions. In both cases, the probability of obtaining a particular result when you make some measurement is obtained from the same mathematical calculation, which is prescribed by the rules of quantum mechanics. It is thus not possible to decide between them (as one does between conflicting theories) by testing them experimentally, because in the absence of a time machine they make the same experimental predictions. It is the interpretation, or way of picturing what is going on, that differs.

According to the Copenhagen interpretation, you calculate the probability that, in some given situation, an observable quantity has a certain definite value, as indicated by the measuring apparatus. In the many worlds interpretation, the observed quantity doesn't have a unique value after the measurement. Instead you are calculating the probability of finding yourself in the particular state or "world" where the measured quantity, as shown by the measuring apparatus, has a certain value. There are, however, other worlds in which other copies of "you" find themselves, in which the measurement has a different outcome. Whichever way you picture it, you wind up with the same probability— that predicted by the rules of quantum mechanics—for observing a given value for the quantity you are measuring.

Since you can't decide on the basis of experiment which interpretation is right, it is basically a matter of taste which one you choose to adopt. For this reason, one might say that few papers have been the subject of more lunch table conversation among physicists than that on the many worlds interpretation. Most physicists probably prefer the Copenhagen interpretation, which is the one we almost always teach our students in their introductory quantum mechanics courses. This avoids the complication of multiple parallel worlds. However, on an intellectual basis, the case can be made that the many worlds interpretation is more internally consistent. In any event, most physicists would agree, perhaps grudgingly, that if you want to adopt the many worlds interpretation, you're at liberty to do so. (Many, however, feel that the necessity of introducing an infinite number of parallel universes just to explain what an electron does is far too much metaphysical baggage.)

Note, however, that the preceding discussion assumes that you can't build a time machine. This assumption, of course, may well be correct. However, we are interested in exploring whether it, in fact, may be possible to build time machines. David Deutsch of Oxford University pointed out in a 1991 *Physical Review* article that if the many worlds interpretation is correct (Deutsch is convinced that it is), an interesting possibility exists. In the case of the grandfather paradox, a time-traveling assassin would discover that he had also arrived in a different Everett "world." Therefore, no paradox would arise when he carried out the dastardly deed.[6]

According to the many worlds interpretation, once you are in a particular world, you are unaware of the existence of the other worlds. Remember our thought experiment above—you're either in a world where the needle points to 1 or to 2. In the former, the electron will be spinning clockwise after the

---

6. Actually, there is one complication we haven't discussed. The full explanation of this is quite technical, and we can only give a brief overview. It turns out that Deutsch's idea requires a significant modification of the rules of quantum mechanics in the presence of a time machine. Instead of describing systems by wave functions, as in standard quantum mechanics, they must be described in terms of what are called "density matrices." These are actually part of the machinery of ordinary quantum mechanics, but there they are used to describe the probable average behavior of a large set of identical systems which do not affect each other. If Deutsch's approach is adopted, one must use a density matrix to describe the behavior of a *single* system. If one attempts to use a wave function, one finds that it undergoes a sudden change—a jump in its value—as one goes through a wormhole. This is discussed in Allen's 2004 *Physical Review* article referred to above. Such discontinuous jumps are unphysical, and in quantum mechanics, such behavior of the wave function indicates that the system being described has infinite energy.

The question as to whether nature is willing to bend the well-established rules of quantum mechanics in order to allow Professor Deutsch's scheme for backward time travel is a question that can't be answered unless we have a time machine so we can do the required experiments. The usual rules of quantum mechanics are very well-established. However, they have never been tested in the presence of a time machine, and one must be cautious about extrapolating physical laws to new situations in which they have not yet been verified.

measurement. You will be unaware of a world where it is spinning counter-clockwise. In that world there will be another copy of "you" who sees the needle of the gauge pointing to 2.

You might ask yourself, "Couldn't I just *push* the needle on the gauge from 1 to 2 and thus find myself in the other world?" The situation is much more complicated than this. The measuring apparatus is a macroscopic device. To describe its state completely, you need to specify not only where the needle points but also its internal coordinates, which are the coordinates of each of the huge numbers of atoms and molecules of which it is composed. So to turn the state of the measuring apparatus in one of the two Everett "worlds" into the state in the other would require readjusting every one of this fantastically large number of coordinates. In other words, not only the macrostates but also all of the microstates of the measuring devices in the two different worlds must be the same. As a practical matter, the probability of this ever happening as the two states of the measuring apparatus evolve over time is so absurdly small as to be effectively zero. Physicists describe this situation by saying the quantum states of the measuring apparatus in the two different worlds are "decoherent."

Now let's consider the grandfather paradox from the point of view of the many worlds theory. We will suppose that we have a time machine in the form of a wormhole like that in figure 9.8, except that now mouth B will be in the year 2260 and mouth A will be in 2200, so the external time difference between the two wormhole mouths is much larger than before. Remember, this doesn't have any connection to the internal length of the wormhole. Therefore, we will still assume that, according to your own watch, very little time elapses for you, the time traveler, between entering mouth B and emerging sixty years earlier from mouth A. Imagine that, for some strange reason, you go back in time to kill your grandfather. In one universe you emerge as an adult from the time machine and kill your grandfather. In that universe you will then go on living out your life. (This may be a short one if your grandfather lived in an era in which capital punishment was prevalent!) In that world you will never be born, so you will never exist as a child or young adult, and hence, you will never enter the time machine. Observers in that world will be somewhat puzzled, because while their records will show that someone emerged from mouth A in 2200, they will not see anyone enter mouth B in 2260, since you entered the wormhole in the other Everett world. This, however, does not constitute a logical contradiction. We do not have the same event both happening and not happening in the *same* world.

In the second Everett world, history unfolds as it already has and as you know of it from your memory or from other records. The past, up to the time you entered the time machine, has already happened in this world and cannot be changed. You will be born, say in 2230, since in this world no homicidal adult version of yourself emerged from the time machine in the past to kill your grandfather. You will then live out your life, as it has already occurred, and eventually enter mouth B of the wormhole in 2260 and disappear, never to be seen again in this world. As before, no logical contradiction occurs in this world, although a puzzling phenomenon will be observed. In this case a time traveler will be seen to enter mouth B in 2260, but there will be no historical record of anyone emerging in this world from mouth A in 2200. Thus, the grandfather paradox has been successfully evaded, exactly as in the many science fiction works based on the idea of "parallel worlds."

Therefore, in the presence of a time machine, the many worlds *interpretation* becomes the many worlds *theory*. The theory could actually be tested by using the time machine to travel backward in time and observing whether you wind up in a new world in which things happen differently than you remember. For example, you might encounter a younger copy of yourself who has not yet entered the time machine. If the theory turns out to be correct, backward time travel without paradoxes would be possible if an advanced civilization figured out how to build a time machine.

Let us briefly mention the way in which information paradoxes are resolved in the many worlds framework. As an example, consider the mathematician's proof paradox, mentioned earlier in the chapter. In the many worlds view, the mathematician receives the proof from a time traveler *who came from a different Everett universe*. In the universe where the time traveler originated, the mathematician became famous by doing the work of proving the theorem himself. The theorem was then published, copied from a textbook by the time traveler, and given to the mathematician in an *alternate* universe. Therefore, the solution to the information paradoxes in the many worlds picture is that the information was generated by normal means, but in a different universe from the one in which the time traveler ends up!

Allen analyzed the many worlds idea in somewhat greater detail in a 2004 article in *Physical Review*, the same journal where Deutsch's paper originally appeared. We take our time machine to be the wormhole in figure 10.2 and again replace human time travelers with billiard balls. We use the model discussed above, which led to a paradox. (As a reminder, we have a billiard ball that is aimed to pass through an open gate then enter mouth B of the wormhole, and emerge from mouth A.) We will refer to this as the "incident ball." We can also call it the younger ball, since it will be younger in terms of the clock carried on the ball itself than it will be if it later goes through the wormhole and then travels back in time. We have a detector positioned outside mouth A of the wormhole that determines if the billiard ball exits the wormhole, and, if it does, sends a radio signal closing the gate so that the ball cannot get to mouth B in the first place.

This is a situation rather like that of the spinning electron, in that there are two possibilities. At a given moment in time, either a billiard ball emerges from mouth A or it does not. We can imagine the ball detector at mouth A has a gauge on it, like the gauge on our electron spin measuring apparatus. In the present case the needle initially points at 0, but turns to point at 1 if a ball emerges from the wormhole. After 4:00 p.m. there will then be two Everett worlds. In one, which we will call the 0-world, observers outside the wormhole will say that no ball has emerged and the needle still points at 0. In the 0-world, since no ball has emerged, no radio signal is sent, the gate remains open, and the incident ball reaches mouth B and enters it at 5:00 p.m. This world corresponds to that depicted in the right-hand side of figure 10.2b (unlike in the left-hand side of figure 10.2b, in this case, no ball is seen to emerge from mouth A). In the other world, the ball emerges from mouth A at 4:00 p.m. and is detected, so the needle turns to point at 1 and a radio signal is sent, causing the gate to close so that the incident ball never reaches mouth B. We will call this the 1-world. This world corresponds to that depicted in figure 10.2c. All of this is just analogous to our discussion of the grandfather paradox. The ball plays your role in our little drama, and the closing of the gate corresponds to your murder of your grandfather.

One might raise the following objection to the many worlds approach. As we have already said, in the usual situation, once a measurement has been made and the branching into different Everett worlds has taken place, those worlds know nothing of one another. Due to the phenomenon of decoherence, it is impossible to go from one to the other. How then can the billiard ball enter the wormhole in the 0-world where the gate is open and wind up in the 1-world where its appearance has caused the needle to point to 1 and the gate to close? The point is that it takes a small, but nonzero, amount of time for the detector to recognize that a ball has emerged and for the needle to turn from 0 to 1. At the instant the ball first emerges, it has not yet been detected and the needle still points at 0 in each of the two Everett worlds. This is the key point. Because it has traveled back in time, the ball starts to emerge slightly before the measurement is completed. In physicist's terminology, the two worlds are not yet decoherent. It is the measurement that causes the sudden branching of the two worlds. At that point, the measuring apparatus recognizes the appearance of the ball in one of the two worlds, and it is naturally in that world that the needle turns to point to 1, and the gate is closed. Thus, once the measurement has been made and the two worlds have branched, the ball, which entered mouth B and then emerged from mouth A of the wormhole in the 0-world, winds up in the world where the needle points to 1. This is exactly the cross connection between the two Everett worlds that Deutsch envisioned. It is possible because the ball, having traveled backward in time, emerges from mouth A of the wormhole slightly before the sudden change in the internal state of the detector that occurs when the emerging ball is observed. It is this sudden change that connects the 0-world to the 1-world.

Let's see how all this looks to different observers. First, what about external observers in the 0-world? They see the incident ball enter the wormhole at mouth B. As far as they are concerned, the ball disappears. No ball emerges from mouth A in this world. The ball is actually not lost but has gone back in time and emerged in the 1-world. Observers in the 0-world, however, will know nothing of this.

Now let's consider what is seen by observers in the 1-world. Observers in this world see a billiard ball emerge from mouth A at 4:00. As a result, the gate closes, and the incident ball is stopped before it reaches mouth B. Thus, external observers in this world will see no ball enter mouth B. Hence, in the 1-world, observers will see a ball appear, seemingly for no reason, out of mouth A. The ball has actually come from the 0-world, but observers in the 1-world know nothing of this.

In the preceding paragraphs, we have described the situation as seen by external observers, that is, those outside the wormhole. Let's also look at things from the point of view of a hypothetical time-traveling intelligent bug, equipped with his own watch, riding on the billiard ball. At 4:00 p.m. on the bug's watch, a branching occurs, and it may wind up in one of two possible Everett worlds, each with a 50/50 chance. In the first one, the bug on the incident ball sees no ball emerge from the wormhole at 4:00 p.m., so the needle of the dial remains pointing at 0, and the gate remains open. This is the copy of the bug that finds himself in the 0-world. Unhindered, it reaches mouth B of the wormhole at 5:00 p.m., as shown on both its watch and the external clocks. The bug enters the wormhole, and emerges from mouth A shortly after

5:00 p.m. on its watch and 4:00 p.m. on the external clocks, since it has traveled one hour back in external time in the brief length of his own time it has taken him to traverse the wormhole. As it emerges, it hears a click as the detector records his presence and sees the reading of the dial change from 0 to 1. It has, in fact, now entered the 1-world, although it feels no sudden change.

Next, the bug notices the gate close and at 4:30 p.m. sees it stop another ball, thus preventing it from reaching mouth B. The other ball has a bug on it that looks very much like itself. However, it has no recollection of ever hitting that gate, which was open when it passed it. But if it should exchange conversation with the other bug, they would discover that their lifetime experiences up until 4:00 p.m. were the same. The 1-world thus contains two copies of the bug. The younger one (in terms of its own time) is the copy of the bug in the 1-world that was initially riding toward mouth B when the two worlds branched at 4:00 p.m. external time. The other is the older (by an hour), according to its own watch. This is the copy of the bug that we have been following up until now. It entered the 1-world after traveling back in time through the wormhole. After encountering its younger self, the older bug then goes off into the future on some trajectory we haven't specified.

The second possibility for our bug is that at 4:00 p.m., as he is heading toward the gate and mouth B, he sees another ball emerge from the wormhole at mouth A. The needle then changes from 0 to 1. It then sees the gate across his path close so that the bug bumps into it at 4:30 p.m., and, let us say, comes to a stop. It then watches the other ball, whose passenger looks very much like a slightly older version of itself, go off on some other trajectory. Again, if it exchanges notes with the other bug he will discover that they lived identical lives up until 4:00 p.m.

Thus, in each of the two worlds, each bug—or it would be more accurate to say each of the two Everett copies of the single initial, pre-4:00 p.m. bug—sees events unfold in a perfectly consistent way. There are no paradoxical contradictions. There is the strange occurrence of encountering itself. However, this is not paradoxical, that is, it involves no *logical* contradiction. The possibility of such an occurrence seems inherent in the idea of backward time travel should that actually be possible.

From what we have said so far, it seems that Deutsch's idea of invoking the many worlds interpretation of quantum mechanics does provide a consistent theory of backward time travel. It also avoids the necessity of seemingly highly improbable occurrences that are the result of the only other such theory, the banana peel mechanism.

"Slicing and Dicing"

The theory as we have presented it until now works very well for indivisible, point-like objects. We believe the electron and various other elementary particles to be examples of such objects. However, we've been ignoring one complication: as discussed in Allen's paper, it appears that, in the many worlds theory, backward time travel may be exceedingly hazardous to one's health. Macroscopic objects, such as people or billiard balls, have many individual constituents, the atoms or molecules of which they are composed. Thus, in principle, they are capable of being broken up into smaller pieces. Such objects take a definite interval of time to exit the wormhole or any other time machine. The front of the billiard ball exits the wormhole before (in terms of external time) its back end does. In the case of a billiard ball the time it takes to exit is just given by the diameter of the ball divided by the speed at which it is moving.

The problem is that, in general, one can build a detector that is sensitive enough to detect the presence of the ball before it has emerged completely. Suppose the time it takes the detector to notice that a billiard ball has appeared is less than the time it takes for the billiard ball to emerge completely from the wormhole. Let's say, for example, that the ball is detected when only slightly more than half of it has emerged. Let us also assume that this detector's sensitivity is such that it will not trigger at all unless slightly more than half of a ball emerges. Then the two Everett worlds we have discussed, the one in which the ball appears out of the wormhole and the one in which it doesn't, will split before the back end of the billiard ball exits the wormhole.

How does this affect our previous discussion? For the first half of the billiard ball, nothing is changed. It emerges from the wormhole when the needle points at 0 (before the two Everett worlds have branched), is detected, and naturally winds up in the 1-world. But when the rear half of the ball emerges, the branching between the two worlds has already occurred. The world in which the rear half emerges, and where the needle points to 0, has lost contact with the 1-world. The two worlds have now become decoherent, and transition between them is impossible. Thus, we have the 1-world containing only half of the ball. In that world the gate has closed, preventing the incident ball from reaching mouth B. However, that world will not contain the second half of the ball, which reached mouth A after the two worlds had branched. It has been left behind in the 0-world. Recall that the second half is actually slightly less than half the ball and is not therefore enough to trigger the detector. Therefore, in the 0-world, the second half of the ball will emerge from the wormhole, but the detector will not be triggered. In this world, the needle will remain pointing to 0, the gate will remain open, and the younger version of the ball will reach the gate at mouth B and go through the wormhole and be split in two. The first half of the billiard ball will wind up in one of the two worlds, the 1-world, and the second half in the other.

In fact, the problem is even more serious. The more sensitive the detection device, the worse the problem becomes. Say we increase the "size sensitivity" of our detector so that it can detect a sliver corresponding to one-fiftieth of a billiard ball emerging from mouth A. Let us also modify it so that, when it detects an emerging sliver, not only does the needle turn to point to 1, but the device also records the time when it made the observation. One would now find 50 different Everett worlds in which a billiard ball sliver had been detected. They would be different worlds, because observers in each one of them would see a different time reading on the dial. In each of them the gate would have been closed as a result of the detection of an emerging sliver, so that the incident billiard ball will not reach the gate at mouth B of the wormhole in any of these worlds. Moreover, each of these worlds will contain only a single sliver of the incident ball.

Thus, if we use a sufficiently sensitive detector, the billiard ball (or for that matter, a person, a spaceship, or anything else traveling back in time through the wormhole) will emerge in a large number of small pieces, each appearing in a corresponding number of Everett worlds.

You might say to yourself, "Well, maybe I can't *personally* survive a trip back in time through a wormhole to take the money in my 401(k) out of the stock market just before the next time Wall Street decides to do something especially stupid and set off a financial crisis. However, I can accomplish the same result by sending myself a warning message backward in time." Unfortunately, however, this strategy runs into the same problem we have just been talking about.

You can model a message containing information as a series of Morse code dots and dashes. If Deutsch is correct, you can send yourself a message containing a single dot. This would be analogous to sending a single electron through the wormhole. However, a message consisting of a single dot doesn't convey much information, particularly since it would be hard to pick your dot out of the random background static that is always present. To convey information would require a message containing a number of dots and dashes. But one would expect such a structured message to have the same problem as an extended material object; the various characters in the message would end up in different Everett worlds. Hence, you would wind up with access to only a single lonesome and uninformative dot that happened to wind up in your particular "world."

Deutsch's idea that the many worlds interpretation of quantum mechanics allows one to avoid the paradoxes of backward time travel provides a very clever way of implementing the "parallel universe" idea of science fiction in a physical context. We are always assuming that the engineering problems of building a time machine have been solved by some advanced civilization (discussion of these problems is provided in chapters 11 and 12). At first sight, the many worlds theory may appear to avoid the paradoxes associated with backward time travel. Unfortunately, as we have seen, it seems to imply that only elementary objects, such as electrons, can survive a trip through a time machine intact. More complicated systems, including human beings, appear to be dissociated into their more elementary constituents in passing through a time machine. This must be true, since the individual constituents are "sliced and diced" into different Everett worlds if a sensitive detector is used to observe the system as it emerges from the time machine. Therefore, we seem to be left with the conclusion that backward time travel by macroscopic systems, for example, people or spaceships, will be possible only through the banana peel mechanism. This means that the laws of physics imply that such time travelers will necessarily encounter a liberal sprinkling of (figuratively speaking) banana peels lying in their path, even though this may seem the result of highly improbable coincidences.

11

"Don't Be So Negative"
Exotic Matter

When your gravity fails, and negativity just
don't pull ya through . . .

BOB DYLAN, "Just Like Tom Thumb's Blues"

You've got to accentuate the positive,
eliminate the negative.

JOHNNY MERCER, "Ac-Cent-Tchu-Ate the Positive"

Negative Energy

We saw in chapter 9 that all of the various mechanisms for time travel and faster-than-light travel involve the use of exotic matter. So what is this stuff? Exotic matter, in the sense that the term is used in this area of physics, is mass/energy that violates the so-called weak energy condition. This condition states that all observers in spacetime must see the local energy density (the energy per unit volume) to be nonnegative. It is a "local" condition in that it is required to hold true at each point in spacetime. All observed types of matter and energy in classical (i.e., nonquantum) physics obey this condition. The reason for postulating such energy conditions in relativity was discussed in chapter 9. Let us briefly summarize the main points.

Given a distribution of physically reasonable matter and energy, one can, in principle, solve Einstein's equations of general relativity to find the geometry of spacetime that arises from this matter and energy. The problem is that Einstein's equations, by themselves, do not tell us what constitutes "physically reasonable" matter or energy. Without some additional assumptions, you can

< 158 >

go the other way. Write down any spacetime geometry that has the properties you want, and from that work backward to find the distribution of matter and energy you need to generate that geometry. If that was really all there was to it, then Einstein's equations would have no predictive power at all. Since any geometry is produced by *some* distribution of matter and energy, you can get anything you like by this procedure. Energy conditions were introduced in relativity as conditions that observable matter and energy seem to obey, which also allow physicists to prove some very powerful mathematical results in general relativity. These include the famous "singularity theorems" of Roger Penrose and Stephen Hawking, which prove the existence of singularities inside of black holes and at the beginning of our universe.

"Negative energy" would be energy (or matter) that violates the weak energy condition. *We will henceforth use the terms negative energy and exotic matter interchangeably.* Let us begin by immediately clearing up a common misconception. Given our discussion in chapter 9, and particularly if you are a *Star Trek* fan, you might think that by exotic matter we mean "antimatter." In *Star Trek*, the starship *Enterprise*'s warp drive is supposedly powered by matter-antimatter reactions. In chapter 9, we told you that exotic matter was required for the Alcubierre warp drive. Ergo, exotic matter must be antimatter, right? WRONG! Let us say—definitively—that exotic matter is *not* antimatter. When a particle and its antiparticle (e.g., an electron and a positron) collide, the result is a shower of gamma rays, which have *positive* energy density. The positron has a charge opposite that of the electron, but both have positive mass. Therefore, when they annihilate the result is positive, not negative, energy. So, *Star Trek* notwithstanding, matter-antimatter reactions will not give us the type of energy needed for warp drive.

When we use the terms "exotic matter" or "negative energy," we also don't think of it in terms of *classical particles* with "negative mass." Suppose for a moment that you could have "negative-mass baseballs." How would they behave? If such objects could exist, the world would be a very strange place indeed. Particles with positive mass are gravitationally attractive; particles with negative mass would be gravitationally repulsive. Since particles with positive mass fall down in the earth's gravitational field, one might think that negative-mass particles should fall up. But that would seem to allow, contrary to what we have seen so far, the possibility of locally distinguishing between gravity and acceleration.

Suppose that we have two rocket ships, one accelerating at a constant rate of 1g in empty space and the other at rest on the surface of the earth. In each rocket ship we have two particles, one with positive mass and one with negative mass. In each rocket ship both particles are released. What will an observer inside each rocket see? In the accelerating rocket, the two particles are released and, as seen by an outside observer, the floor accelerates up to meet them. An observer *inside* the rocket will see both particles "fall" downward to the floor of the rocket at the same rate, that is, with an acceleration of 1g. For the rocket on the surface of the earth, when the particles are released, the positive-mass particle will fall downward and, according to our reasoning above, the negative-mass particle should "fall" *upward*. But that means the two observers inside the rocket will see different situations, even though, according to the principle of equivalence, they should see the same thing. Would a negative-mass particle really fall upward in the earth's gravitational field? (This is one that has generated a bit of confusion even for some famous physicists.)

Let's examine the situation more carefully. We will denote the mass of the positive-mass particle m, and that of the negative-mass particle $-m$ (where we assume that $m > 0$). The scenario described above contains a hidden assumption. In drawing our conclusion, we implicitly assumed for the negative-mass particle that, although its gravitational mass was negative, its *inertial* mass was positive. Let us instead assume that the principle of equivalence holds for the negative-mass as well, including for the *sign* of the mass. That is, if the gravitational mass of a particle is negative, then so is its inertial mass. Then, as we will show, the seemingly paradoxical situation described above can be resolved.

Using Newton's law of gravitation, we can determine the direction of the earth's gravitational force on each particle. Recall that the law of gravitation is

$$F = -\frac{GmM}{r^2},$$

where M here is assumed to be the mass of the earth, G is Newton's gravitational constant, and r is the distance of the particle from (the center of) the earth. For the positive-mass particle, the force is just given by the equation above; the minus sign indicates that the direction of the force is downward. For the negative-mass particle we have $F = -G(-m)M/r^2 = +GmM/r^2$; the plus sign indicates that the sign of the gravitational force is upward. But to find the direction of the acceleration of the particle we have to use Newton's second law of motion: $F = ma$. For the positive-mass particle, we have $F = +ma$, and $F = -GmM/r^2$. Setting these two expressions equal and canceling the m's, we get $a = -GM/r^2$, so the direction of the acceleration is downward, as we would expect. This is because, for a positive-mass particle, the acceleration is in the same direction as the force. For the negative-mass particle, we have $F = -ma$ (since the mass of the particle is $-m$), and $F = +GmM/r^2$. Setting the two expressions for F equal and cancelling the m's yields $a = -GM/r^2$, so in fact the negative-mass particle *also* accelerates *downward*! This is because, for the negative-mass particle, even though the force is directed upward, the force and the acceleration are *oppositely* directed, unlike for the positive-mass particle where they are in the same direction. So therefore one cannot use negative-mass particles to locally distinguish between gravity and acceleration.

Suppose that we had a negative-mass planet of mass $-M$ instead, and we release a small positive-mass particle near it. What would happen? The gravitational force between the two is repulsive ($F = +GmM/r^2$). So the direction of the gravitational force on the positive mass is upward. For a positive mass, the force and the acceleration are in the same direction($F = +ma$). Therefore, the positive mass would be repelled from the negative-mass planet. A small negative-mass particle released near the planet would feel a gravitational force directed toward the planet. This is because in this case we have $F = -GmM/r^2$, since the two minus signs in front of the masses cancel each other out. However, since for the negative-mass particle, force and acceleration are oppositely directed ($F = -ma$), the negative-mass particle would also accelerate *away* from the planet!

Now let's look at a more unusual situation. Consider a positive- and a negative-mass particle, with masses m and $-m$, respectively, out in space far from all other gravitating bodies. If the two particles are released from rest what will they do? This situation is shown in figure 11.1. The negative-mass

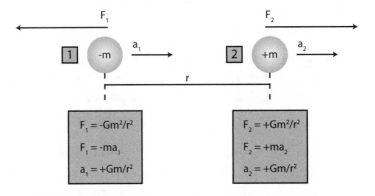

FIG. 11.1. Positive and negative masses. The negative mass will chase the positive mass!

particle is labeled 1, and F_1 is the force particle 1 experiences due to particle 2; similarly, a_1 is the acceleration experienced by particle 1. The same conventions apply for particle 2. (The positive direction of force and acceleration is taken to be to the right in the diagram.)

Figure 11.1 shows that the positive-mass particle is repelled by the negative-mass particle and accelerates away from it. The negative-mass particle is also repelled from the positive-mass particle, but because for it the force and the acceleration are oppositely directed, it accelerates *toward* the positive-mass particle. The net result is that the two particles chase each other, while maintaining a constant distance from one another, with ever-increasing speed! In fact, the same thing would happen in the case of the positive-mass particle suspended above the negative-mass planet discussed above. There the repulsive downward force on the planet would cause the planet to accelerate upward. But, because of the large mass of the planet and Newton's second law in the form $a = F/m$, the acceleration of the planet would be too small to notice.

Now at first sight, this would seem to be a flagrant violation of the laws of conservation of energy and momentum. In this context, the law of conservation of energy would say that the kinetic energy (energy of motion) + potential energy (energy of position) of the two-particle system remains constant, since the particles constitute an isolated system. Similarly the law of conservation of momentum would say that the sum of the mass times the velocity (speed and direction) for the two particles remains constant. The kinetic energy (in Newtonian physics) of a particle is $KE = \frac{1}{2}mv^2$, where v is the speed of the mass m. The momentum of a particle is its mass times its velocity, $p = mv$, where, in this equation, the *direction* of motion must also be taken into account. So the law of conservation of energy for our two-particle system reads:

$$\text{total kinetic energy} = KE_1 + KE_2 = \frac{1}{2}(-m)v^2 + \frac{1}{2}mv^2 = 0.$$

(Here we have ignored the potential energy, since the potential energy of each particle remains constant because their *relative* positions remain constant.) Similarly the law of conservation of momentum would be:

$$\text{total momentum} = p_1 + p_2 = (-m)v + mv = 0.$$

So here we have a very peculiar situation where the laws of conservation of energy and momentum are obeyed—in fact, the total kinetic energy and total momentum of the system are both zero—yet the two particles chase each other with ever-increasing speeds! One cannot help the feeling that there is still something wrong with this scenario, since it seems to be an example of a "free lunch," one that we don't see occurring in the real world. In passing, we note that if someone threw a negative-mass baseball at you, because the force and acceleration are oppositely directed, you would have to hit it in the direction it was moving in order to stop it![1] As far as we know, all particles in our world have positive mass.

We now turn to the concept of negative energy as described by the laws of quantum mechanics. One of the most amazing discoveries of twentieth-century physics is that what we normally consider empty space, the "vacuum," is not really empty at all! The laws of quantum mechanics have taught us that the vacuum can be described as a roiling sea of "virtual particles," or, "vacuum fluctuations": particles that appear out of and disappear back into the vacuum so rapidly that they cannot be directly measured. This modern picture of the vacuum is a consequence of the "energy-time uncertainty principle" proposed by Werner Heisenberg in the early 1920s. To measure the energy of a system to within a certain accuracy ΔE, takes a certain amount of time; call this time interval ΔT. To probe ever-smaller regions of space requires measurements of ever-shorter duration, that is, ever-smaller values of ΔT. However, according to the energy-time uncertainty principle, the shorter the time duration of our measurement, the larger the uncertainty in the energy of the system being measured. The product of the two can never be smaller than a certain universal constant of nature, that is, $\Delta E \, \Delta T \geq \hbar$, where $\hbar$ is Planck's constant divided by 2π. Planck's constant, named after Max Planck, one of the founders of quantum mechanics at the turn of the twentieth century, is a universal constant of nature like the speed of light. It governs the scale of the very small, much like the speed of light determines the scale of the very fast. Planck's constant is a very small number, when expressed in terms of ordinary "everyday" units such as kilograms, meters, and seconds. This is why we don't notice quantum effects on the everyday scale of things. As a result of the energy-time uncertainty principle, if one is measuring the energy contained in a region of space over a very short timescale, the uncertainty in the energy measurement will be very large. This uncertainty makes it possible for particles to appear and disappear from the vacuum over this timescale without being directly observed.

We said that virtual particles, or vacuum fluctuations, occur so rapidly that they cannot be directly detected. Well, you say, I thought physics deals with

1. For a more extensive discussion of negative mass in classical physics, see the delightful article by Richard Price, "Negative Mass Can Be Positively Amusing," *American Journal of Physics* 61 (1993): 216–17.

things you can measure! Yes it does. It turns out that the indirect effects of vacuum fluctuations are measurable. An example is the Lamb shift in the spectrum of hydrogen (named after Willis Lamb, who was the first to measure it). The "spectral lines" (specific wavelengths of light given off by atoms, which are unique to individual chemical elements) are slightly shifted from where they were expected to be. This difference in the position of the lines can be shown to be due to vacuum fluctuations.

We shall discuss several examples of negative energy in quantum physics. The first is the "Casimir effect," discovered by Hendrik Casimir in 1948. He predicted that two uncharged parallel metal plates, when placed close together, would experience an attractive force due to vacuum fluctuations. This force has been measured on several occasions, with the most recent measurements agreeing with Casimir's prediction to within a few percent. From the expression for the force, one can calculate the energy density between the plates. Remarkably—and very germane to our purposes—this energy density turns out to be negative. That is, the energy density between the plates is lower than that of the vacuum when the plates are not present. The energy density varies as $-1/d^4$, where d is the distance between the plates, which means that the amount of negative energy increases the closer the plates are together. Although the force between the plates has been measured, the energy density is far too small to measure directly. We see that to get a large negative energy in the Casimir effect, it has to be confined to a very thin region between the plates.

A second effect is the "evaporation" of black holes, predicted by Stephen Hawking in 1975. Hawking showed that when quantum field theory (i.e., the laws of quantum mechanics applied to fields) was applied to black holes, it predicted that particles and radiation could "leak out" of the hole, reducing its mass in the process. There are a number of ways to think of this process. One is that the presence of the black hole disturbs the vacuum around it, causing a flow of negative energy into the hole, which pays for the positive-energy "Hawking radiation" that a distant observer sees by decreasing the mass of the black hole. The rate of radiation depends on the inverse fourth power of the mass of the hole. As a result, this effect is very tiny for stellar mass black holes. However, it is possible that there could exist "mini" black holes, with about the mass of a mountain and the size of an elementary particle, which might have formed in the very dense early universe. For such mini-holes, the rate of Hawking radiation is very large, causing them to explode violently. Even though Hawking radiation has not been observed (yet), and although the effect is tiny for black holes that we are likely to encounter, Hawking's work is of profound importance. His result made the laws of black hole physics consistent with the laws of thermodynamics and has deep implications for three areas of physics: general relativity, quantum theory, and thermodynamics. And negative energy plays a crucial role in making this possible.

A third illustrative example of negative energy in quantum theory is the class of quantum states known as "squeezed states" of light. In a classical electromagnetic wave, the electric (or magnetic) field is well-defined at each point in time and space. But in quantum mechanics, because of the quantum fluctuations mandated by the uncertainty principle, the field can only be approximately localized in space and time. For example, draw a sinusoidal curve on a piece of paper. This might represent a classical electric field that is varying with time at some point in space. The quantum electric field has random fluctuations as a function of time superimposed on the regular classical time variation and so would be depicted as a blurry sinusoidal curve, compared to the sharp curve for the classical electric field.

However, one can "cheat" the uncertainty principle in the sense that one can decrease the fluctuations in one characteristic of the wave, say, the phase, below the uncertainty principle limit while increasing them in another feature, say, the amplitude. (The amplitude measures the height of the wave crests, and we can think of the phase as determining when, along the time axis, the wave crests occur. More precisely, the phase determines the value of the electric field at $t = 0$. Two waves whose wave crests occur at different times are said to be "out of phase.") This is illustrated in figure 11.2. Of course, one is not really

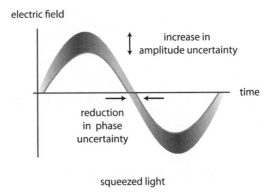

FIG. 11.2. Squeezed light. The uncertainty in the phase ("position") of the light wave is reduced at the expense of increasing the uncertainty in its amplitude (height of wave crests).

circumventing the uncertainty principle, but rather, "redistributing" the quantum fluctuations from one variable to another.

The vacuum state is classically the state with no electric field, but the quantum vacuum has fluctuations in it and is therefore "smeared out" around the zero value for the electric field. One can also "squeeze" the quantum vacuum to create a so-called squeezed vacuum state. As a simple analogy, think of the quantum vacuum as a long water balloon (that should really have a somewhat "fuzzy," i.e., ill-defined surface, like our blurry quantum electric field wave discussed earlier). The creation of a squeezed vacuum state is analogous to squeezing the water balloon at various places along its length. At those places, the squeezing makes the balloon thinner at the expense of making it thicker elsewhere along its length. Squeezing the quantum vacuum decreases the vacuum fluctuations at some places and increases them in others. Squeezed vacuum states are now routinely produced in quantum optics labs and have technological applications that range from the reduction of noise in gravitational wave detectors to the creation of more efficient quantum information processing algorithms.

For our purposes, squeezed vacuum states are interesting because they are states that involve negative energy. Figure 11.3 shows a sketch of the energy density in a squeezed vacuum state as a function of time. We see that there are periodic "pockets" of negative energy density surrounded by regions of (larger) positive energy density. Although these states have been produced in

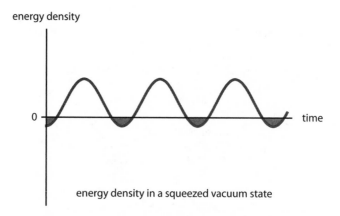

FIG. 11.3. Energy density in a squeezed vacuum state. A squeezed vacuum state has oscillating regions of positive and negative energy. Note that the positive regions are always larger than the negative regions.

the laboratory, as in the case of the Casimir effect, the negative energy density in these states is far too small to directly measure.

In the 1960s, Epstein, Glaser, and Jaffe proved mathematically that any quantum field theory contains quantum states for which the energy density can be negative at a point. Their argument was later generalized to show that one could also find states where the energy density is *arbitrarily* negative at a point. So quantum field theory inherently allows violations of the weak energy condition.

So we have seen that quantum theory forces us to take the idea of negative energy seriously. On the other hand, if the laws of physics place no constraints on negative energy, all sorts of things might become possible. Some of these include wormholes and warp drives, time machines, violations of the second law of thermodynamics (e.g., refrigerators without power sources), and the destruction of black holes. These might be good or bad, depending on one's point of view.

Averaged Energy Conditions

In the case of the energy conditions, theoretical physicists realized that if these constraints are true, then one could prove some very powerful results, such as the existence of the big bang in which our universe began and the formation of singularities at the centers of black holes. It also happened that these conditions, in addition to appearing very reasonable, were also actually satisfied experimentally by classical forms of matter and energy. However, it was then later realized that quantum matter and energy could violate all of the known energy conditions. What to do?

As physicists, we essentially play a game with nature. We make guesses—hypotheses—about how we think the world behaves. These guesses can be motivated by a variety of reasons. One might be, "Gee, wouldn't it be neat if the world was really like this?" Or, "If this is true, then I can prove that a number of these *other* very interesting things must also be true." Sometimes the motivation goes the opposite way: "If this is true, the world would be completely crazy, and things would be happening that we don't actually observe." In cases like the latter, we are motivated to try to understand the reason why the world *doesn't* behave in this or that crazy way. The ultimate arbiter of our guesses must be confrontation with the real world, that is, observation and experiment. Theoretical physicists build simplified models of the world that they hope capture its central features but that are tractable enough to make predictions. The experimental physicists check how successful or unsuccessful the theorists have been.

Given the important role energy conditions play in a variety of areas in relativity, it is imperative to see whether there are weaker constraints on negative energy than the energy conditions, which we know are violated. It may be that one can find such conditions that allow you to preserve your previous results, but that real fields don't obey even those weaker restrictions. The argument of a theoretical physicist would go something like this: (1) "Show that if condition A is true, then we can prove that statement B is true"; (2) "Is there any reason to believe that condition A is actually satisfied in the real world?" As for the energy conditions, we think: "What other weaker assumptions would allow us to prove some of the same results, and yet at the same time might have a chance of being true?" The usual conditions are violated at a point in spacetime, but one never really measures something at a single point in spacetime. Measurements are made over regions of space and take some minimum amount of time. With this in mind, one possibility is that, while quantum field theory allows energy conditions to be locally violated (e.g., at a point or in a limited region), it could be that a suitable average of the energy density, say, over an observer's worldline, is always nonnegative. "Averaged energy conditions" were first introduced by Frank Tipler (now at Tulane University) in the 1970s. He showed that many of the known results in general relativity could be proven using these weaker conditions.

Various forms of an averaged weak energy condition were proposed by a number of people, first by Greg Galloway at the University of Miami, and also by Arvind Borde at Long Island University, and by Tom. This type of condition averages the energy density along the worldline of a geodesic (i.e., freely falling) observer. Such a condition, if true, would say that, although an observer might encounter negative energy at some point along his worldline, he would also have to see a compensating amount of positive energy either before or afterwards. A related, although independent, condition that has played a very important role in this area of research is the "averaged null energy condition." Loosely speaking, it is like the averaged weak energy condition, except that the average is taken over a null or lightlike geodesic. This condition will play an important role in our discussions. John Friedman (at the University of Wisconsin–Milwaukee), together with Kristen Schleich and Don Witt (both at the University of British Columbia) proved the so-called topological censorship theorem, which shows quite generally that the maintenance of traversable wormholes generically requires violations of the averaged null energy condition. As we will see later, Stephen Hawking showed that violations of this condition are also required to build time machines in finite regions of spacetime.

Although the averaged weak and null energy conditions are satisfied over a large range of circumstances, they are known to be violated in others. The averaged weak energy condition holds in flat spacetime when there are no boundaries (e.g., like Casimir plates). The averaged null energy condition holds in flat spacetimes and has been recently shown to hold even in flat spacetimes with boundaries. However, both conditions, as currently formulated, fail in some curved spacetimes. One problem is that even if such conditions were true, it would still leave a lot of wiggle room for creating mischief with negative energy. For example, suppose an observer's worldline initially takes him through a region of negative energy. The averaged weak energy condition would say that he must subsequently encounter a region of compensating positive energy, but it does not specify any time frame for *when* that positive energy must arrive. Suppose the positive energy does not arrive for, say, 25 years. That's 25 years during which time the observer could conceivably manipulate the negative energy to produce large effects, for example, to violate the second law of thermodynamics.

Quantum Inequalities

There has been a parallel—and closely related—line of research to the study of averaged energy conditions. Again, the idea is to see whether there are restrictions on negative energy density *over an extended region*, such as an observer's worldline, as opposed to at a single point.[2] Such constraints, if they exist, might allow the violations of the energy conditions that we know of but still be strong enough to prevent all hell from breaking loose. This program was initiated by Larry Ford in 1978. He suggested that quantum field theory might place bounds on negative energy "fluxes" (that is, flows of energy) and densities that have a form similar to Heisenberg's energy-time principle relation, but with the inequality sign going in the opposite direction: $|\Delta E| \Delta T \leq \hbar$, where $|\Delta E|$ is the magnitude (i.e., the absolute value) of the negative energy, ΔT is its duration, and $\hbar$ is again Planck's constant divided by 2π. Figure 11.4 is meant

2. The quantum inequalities we discuss here are averages over an observer's *worldline*. Very recently, Chris Fewster and Calvin Smith (then also at York) have proven some quantum inequalities that hold over volumes of space and time, not just along worldlines. This is an ongoing area of research.

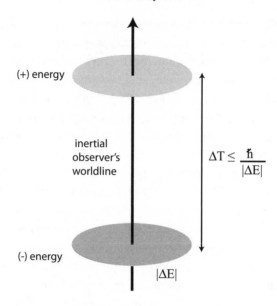

(+) energy

inertial
observer's
worldline

(-) energy

$\Delta T \leq \dfrac{\hbar}{|\Delta E|}$

$|\Delta E|$

FIG. 11.4. A depiction of the physical interpretation of
a quantum inequality.

to give a flavor of the physical implications of such a bound. Consider an observer who initially passes through an amount of negative energy given by $|\Delta E|$. The bound says that not only must compensating energy be encountered later but that it must be encountered no later than a time $\Delta T \leq \hbar/|\Delta E|$ afterward. The implication of the constraint is that the *larger* the magnitude of the initial negative energy through which the observer passes, the *shorter* the time interval before the positive energy arrives.

Ford showed that this kind of bound was obeyed by a certain limited class of quantum states, and conjectured that it might be true in general. In 1991, Ford gave a formal proof that this type of bound held for negative energy fluxes in *arbitrary* quantum states in certain quantum field theories, an extremely powerful result. In 1995, Ford and Tom generalized his proof to include negative energy density. These bounds have since come to be called "quantum inequalities." In light of their implications for wormholes and warp drives, it is worth discussing the (more precise) form of the bounds in a little more detail.

Here we shall discuss the form of quantum inequalities that applies to what are called "free fields," such as the electromagnetic field in flat spacetime. These lead to simplified theories in which we ignore the interactions between fields that occur in nature. The general case of "interacting field theories" is more difficult to deal with mathematically.

In flat spacetime (no gravity), suppose that we have a quantum field that has a region of negative energy density and an *arbitrary* inertial (constant velocity) observer whose worldline passes through this region. Imagine that the observer has a device that "samples" the energy density over some timescale, τ_0, called the "sampling time." Mathematically, we represent this process by what's called a "sampling function." For the purposes of this discussion, just think of the sampling function as analogous to a measuring device that takes some time to switch on and off, and that does most of its measuring over a time τ_0. We can choose τ_0 to be anything we like. Let us call the magnitude of this "sampled energy density," $|\bar{\rho}|$. Then the quantum inequality for energy *density* has the form:

$$|\bar{\rho}| \leq \frac{C\hbar}{c^3 t_0^4},$$

where c is the speed of light, and C is a constant, typically much smaller than 1, but whose numerical value depends on the particular shape of the sampling function. The sampling function is required to be "smooth," that is, no jumps or kinks, and generally mathematically well behaved. There are lots of possible functions that have this form. Each one might represent a slightly different type of measuring device. For example, one sampling function might represent a device that takes 5 seconds to gradually reach full sensitivity, makes its measurement over an average time of 10 seconds, and takes another 5 seconds to gradually shut off. Another might represent a different device that reaches maximum sensitivity in $\frac{1}{10}$ second, measures over an average time of $\frac{1}{100}$ second, and shuts off gradually after an additional $\frac{1}{10}$ second.

Let's examine the right-hand side of the quantum inequality. Now, $\hbar$ is a very small number in everyday units and appears in the numerator, and c^3 is a very big number that appears in the denominator, and C is a constant that is small, compared to 1. Therefore $C\hbar/c^3$ is a very small number. In principle we can choose the sampling time, τ_0, to be anything we want. The choice of a sampling time that yields a strong bound is a bit like Goldilocks's selection of the best bowl of porridge. If the sampling time is taken to be very short, then our sampling function could, for example, be nonzero in the middle of the region of negative energy, which might be very negative there but then rapidly drops off. So this choice gives us a very weak bound. A sampling time that is too long samples the positive energy as well, and as a result does not probe the negative energy optimally. The sampling time that is the same as the duration of the negative energy does the best job.

Therefore, to get a strong bound on the negative energy, we want to choose τ_0 to be *the time over which the negative energy density lasts*. So we see that $|\bar{\rho}| \leq$ (a very small number) / τ_0^4. We see that the larger τ_0 is, that is, the time over which the negative energy density lasts, the smaller (in magnitude) the negative energy density. If we want to go the other way, and make the negative energy density very large in magnitude, then the quantum inequality bound says that the negative energy cannot last very long.[3] If we have regions of negative energy and positive energy density that are separated in time, then we might want to choose the sampling time to be equal to that time separation. In that case we would get a bound on the time it takes for the positive energy to arrive, given some initial amount of negative energy, which is more like the case illustrated in figure 11.4. If we let the sampling time go to zero, then we are "sampling" only over a single point, and the energy density can be arbitrarily negative at a single point in spacetime, which is consistent with earlier results. If we let the sampling time become infinitely long, then we sample over the observer's entire worldline, and our quantum inequality bound reduces to the averaged weak energy condition. So in flat spacetime, the averaged weak energy condition follows from the quantum inequality bound.

The power of the quantum inequalities is that they are proven to hold for *all* quantum states and all inertial observers (in flat spacetime with no boundaries—the Casimir effect will be discussed separately). This includes the squeezed vacuum states discussed earlier in the chapter. One could get at least a hint that this might be true from figure 11.3, where the positive energy peaks are seen to "outweigh" the negative energy troughs.

Another point is also worth emphasizing. Although the quantum inequalities bear a resemblance to the energy-time uncertainty principle, the latter principle was *not* assumed in their derivation. These are rigorous mathematical bounds that are derived directly from quantum field theory, and they have been derived using several different mathematical methods. Were the quantum inequalities found experimentally to be incorrect, then there would be something deeply wrong with the structure of quantum field theory—a theory that has withstood thousands of laboratory tests.

The original quantum inequality bound on energy density derived by Ford

---

3. There is no restriction on the maximum amount of positive energy one can have.

and Tom assumed a specific choice of sampling function. A few years later, Chris Fewster and Simon Eveson, at the University of York in the United Kingdom, gave a much simpler derivation of the quantum inequalities. In contrast to the Ford-Roman analysis, which used a particular sampling function, Fewster and Eveson's method had the additional bonus of being applicable for *arbitrary* sampling functions (again, assuming the functions are smooth and otherwise mathematically well-behaved).[4] Aside from Larry Ford, Chris Fewster, together with his students and collaborators, has probably contributed more to this field than anyone else. His highly rigorous and powerful mathematical techniques have allowed generalizations of the original quantum inequalities, including some that hold in curved, as well as flat, spacetime. Quantum inequalities have been proven for fields that we know exist in nature, such as the electromagnetic field and the quantum field for the electron (the so-called Dirac field), as well as for a number of other fields that may exist. With one exception, however, they have only been proven for free field theories. To summarize all the results on quantum inequalities over the last twenty years would require another book.

All Good Physics Is Done with Mirrors

The quantum inequalities do not forbid the existence of negative energy, per se; what they highly constrain is the *arbitrary separation* of negative and positive energy. Otherwise, one might imagine taking a beam of radiation containing regions of negative and positive energy, somehow splitting off the positive energy and directing it to some distant part of the universe, and bringing the isolated negative energy back to your lab. Paul Davies proposed just such a scenario in his book *How to Build a Time Machine* (2001). As Davies and Stephen Fulling, then both at King's College, London, showed theoretically in the 1970s, one could produce pulses of positive and negative energy by varying

---

4. It should be noted that every sampling function gives a *true* bound on the energy density, but not necessarily the *best* bound. A poor choice of sampling function may give a very weak bound, whereas a more judicious choice might give a much better bound. Both bounds are true, but one provides a stronger constraint than the other. For example, let's say that the amount of money we can afford to spend on a car in a year is represented by x. If our yearly income is $50,000, then obviously $x \leq \$50,000$. Suppose we then determine that the amount which we can set aside per month out of our salary, after deducting our other monthly expenses, is a maximum of $1,000. That gives us a bound of $x \leq \$12,000$. Both bounds are true; however, the latter tells us more about the possible value of x than does the former.

the acceleration of a moving mirror. In practice, the amount of radiation is exceedingly small, unless the accelerations involved are enormous. Nevertheless, one could consider the following scenario. Produce an initial pulse of negative energy, followed by a period of no radiation, and then a subsequent pulse of positive energy (there are mirror trajectories that do this). Use a second mirror (or set of mirrors) to reflect the negative energy in one direction. During the time separation between the pulses, slightly rotate the second mirror to a new position so that when the pulse of positive energy arrives, it gets deflected off at a slightly different angle. Far away from the second mirror, the pulses of negative and positive energy will get farther and farther apart. Do this repeatedly to obtain a large amount of isolated negative energy, which you can then use to build your wormhole, warp drive, time machine, or what have you.

However, consider an inertial observer who is very far away from any of the mirrors and whose worldline intersects the negative energy. Since the positive energy has been deflected somewhere else, this observer would encounter only negative energy, with no compensating positive energy. But this would violate the quantum inequalities, which hold for all quantum states, however they might be produced, and for all inertial observers in flat spacetime. So this scenario is ruled out by the quantum inequalities. What must presumably happen is that in the process of rotating the mirror, positive energy is produced (as we just said, moving mirrors can produce as well as reflect positive and negative energy) that compensates the negative energy. The distant observer would then have to encounter both negative and compensating positive energy. The same would be true if one tried to isolate the pulses by capturing the negative energy in a mirrored box. If one tries to close the door of the box before the positive energy pulse arrives, the closing door acts like a moving mirror that produces compensating positive energy.

As an aside, we mention that in two-dimensional spacetime (one time and one space dimension), where the Davies-Fulling analysis was actually performed, there is only one space dimension for the mirror to move in. As a result, it turns out that a mirror that emits an initial isolated pulse of negative energy must necessarily hit any inertial observer who intercepts the pulse, unless it is stopped before collision. The act of stopping the mirror produces an (even bigger) pulse of positive energy. One can show that, in the latter case, for any inertial observer who intercepts the first pulse, the time interval between the pulses, ΔT, and the size of the negative energy pulse, $: |\Delta E|$, are constrained by the relation $: |\Delta E| \Delta T \leq \hbar$. Hence, the larger the initial negative energy pulse, the shorter the time interval before the positive energy arrives.

(In four dimensions, the problem is much more complicated, and it is difficult to get general exact solutions, which is why Davies and Fulling chose to work in two-dimensional spacetime. This is an example of the situation mentioned in chapter 6, where one may be able to get valuable insight into a complicated four-dimensional situation from a two-dimensional "toy" model.

Quantum Interest and the Casimir Effect, Again

Another example of the richness of the quantum inequalities is that they predict an effect known as "quantum interest." If we consider negative energy to be an energy "loan," then it turns out that nature is a shrewd banker. Not only must the loan be "repaid" with positive energy within a certain limited time period, as we have already seen, but, as it turns out, the positive energy must *overcompensate* the negative energy. That is, the loan must be "repaid with interest." Furthermore, the amount of overcompensation increases with the magnitude and duration of the debt. For example, suppose that we are initially given some fixed amount of negative energy. The longer one staves off the arrival of the subsequent positive energy, within the time limit set by the quantum inequalities, the larger the amount of positive energy must be when it arrives.

Let us return to the case of the Casimir effect, which is a counterexample to the averaged weak energy condition and the quantum inequalities as discussed so far, since the negative energy between the plates doesn't depend on time, and therefore can be made to last as long as one likes. However, we also saw that the magnitude of the negative energy varies as one over the fourth power of the distance between the plates. This means that to get a large negative energy density it has to be confined to a very thin region of space, that is, the distance between the plates has to be very small. (Of course, the area of the plates can, in principle, be made as large as we like.) However, in practice there is a limit on how close the plates can realistically get to each other. The calculation of the Casimir energy density assumes that we can treat the plates as approximately smooth and continuous, that is, we ignore the fact that the plates are really made of individual atoms. Once the distance between the plates is roughly about the size of the distance between atoms in the plates, then this approximation breaks down. At this point we would not expect the energy density to be accurately modeled by the Casimir expression. This means that we cannot use the Casimir effect to generate *arbitrarily* large negative energy densities in the lab.

We might ask, is there another way to "beef up" the negative energy between the plates, for example, by changing the quantum state of the field? In other words, by changing the quantum state of the field between the plates to one different from the usual "Casimir vacuum" state, can one depress the negative energy *below* that of the Casimir vacuum while keeping the distance between the plates fixed? It turns out the answer is *no*. If we take the difference in energy density between the Casimir vacuum state and in any other state, we find that this *difference* obeys a quantum inequality. The implications of this "difference inequality" are that one cannot reduce the energy density below that of the Casimir vacuum energy density by an arbitrarily large amount for an arbitrarily long time. So to make the energy density more negative and static with time, one is forced to confine the negative energy to a narrower region of space.

On the other hand, Ken Olum and Noah Graham at Middlebury College have constructed an example of two interacting fields in flat spacetime, one of which models a confining region (analogous to the Casimir plates) and the other a field confined within that region. They found that, as in the Casimir effect, one could get regions of *static* negative energy that could be maintained as long as one liked. The averaged weak energy condition does not hold in this model (nor does it for the Casimir effect), because one can always choose to average over the worldline of an observer who just sits in the static negative energy region forever. Unlike the Casimir case, here it is conceptually a bit more difficult to justify a difference inequality–type argument, because there is not as clean a separation between what constitutes the confining wall and what constitutes the field being confined. It should be mentioned that their model is simplified in the sense that it involves only two spatial dimensions instead of the usual three. Olum and Graham used that assumption to make the calculation tractable. However, their result does imply that the quantum inequalities do not hold (at least in their original form) for some interacting fields.

It should be pointed out that the Olum-Graham example is similar to the Casimir effect in an important respect. Their model likely has the property that the magnitude of the negative energy density and its spatial extent are inversely related, that is, large negative energies are confined to narrow regions. Furthermore, there is a larger region of positive energy just outside the negative energy, so you cannot have *isolated* negative energy. So their example does *not* show that you can have a lump of negative energy without some positive energy close by.[5]

---

5. However, since the situation they analyze is the lowest energy state of their system, that state is analogous to the Casimir vacuum state. Therefore, if one considers the difference in energy

Interestingly, Olum and Graham showed that in their model the averaged null energy condition is obeyed. This is because any light ray that passes
through the region where the energy density is negative must also pass through
nearby regions of very large positive energy. As a result, the type of energy in
their model is unlikely to be useful for building wormholes, since *violation of*
the averaged null energy condition is a necessary requirement for traversable
wormholes.

The same is true for the Casimir effect. In that case, it turns out (for technical reasons) that the local null energy condition is obeyed for light rays that
move between but parallel to the plates. But the condition is violated along
light rays moving between the plates in the direction *perpendicular* to them.
However, for the *averaged* null energy condition, we have to average over the
entire path of the light ray, which includes the parts that intersect the plates.
The positive energy of the plates more than offsets the negative vacuum energy
between the plates. One dodge we might consider is to drill tiny holes through
the plates to allow the light ray to pass through without intersecting the plates,
so that the light ray encounters only the negative energy. Remarkably, as Graham and Olum showed in a later paper, the contribution to the average due to
effects caused by the presence of edges of the holes in the plates is *positive* and
outweighs the contribution from the negative energy between the plates! This
result has been recently extended to include general types of boundaries, in
work done by Fewster, Olum, and Mitch Pfenning (a civilian faculty member at
the United States Military Academy at West Point).

This result is important because it shows that one cannot use the Casimir
energy to maintain a traversable wormhole. In the original wormhole time machine model proposed by Morris, Thorne, and Yurtsever, they placed a pair of
Casimir plates near the wormhole throat and used the Casimir vacuum energy
to provide the required negative energy to hold the wormhole open. However,
in this model, for an observer to get through the wormhole, she has to pass
through the plates. If we imagine cutting holes in the plates to allow her to pass
through, we ruin the delicate negative energy balance that holds the wormhole
open (as the results discussed in the previous paragraph show).

Although the original form of the averaged null energy condition can be
violated in some spacetimes, it seems at least possible that a suitably modified
version—one that has the same physical implications—holds. These implica-

---

between that state and any other quantum state of the system, it is quite likely that one could prove
"difference inequalities" for this system as well.

tions would include no traversable wormholes and no time machines. On the one hand, evidence for this is supplied by the work done by Fewster, Olum, and Pfenning, as well as more recent work by Olum and Graham.

However, some very new work by Doug Urban (also at Tufts) and Ken Olum shows that one can always set up some situations where the averaged null energy condition is violated. Our aim in mentioning this work is primarily for completeness and to point out that this remains an ongoing area of research. Suppose you have a lightlike geodesic and along some parts of the geodesic there is negative energy. Their proof involves the use of what is mathematically called a "conformal transformation" along the lightlike geodesic, which essentially enhances the magnitude of negative energy regions to the average along the geodesic. The technical details are far beyond the scope of our discussion here. Whether these newly discovered violations of the averaged null energy condition are applicable in wormhole and warp drive spacetimes is not yet known.

Urban and Olum point out that the various formulations of the averaged null energy conditions are all for "test fields," that is, fields that are weak enough so as to not alter the background spacetime geometry. (Think of the analog of a tiny marble rolling on a rubber sheet with a huge bowling ball sitting in the middle of the sheet. The marble is a "test particle" in the sense that its contribution to the curvature of the rubber sheet is negligible compared to that of the bowling ball. A "test field" is the same idea.) They speculate that a "self-consistent" averaged null energy condition, that is, one that properly takes into account the gravitational effects of the test field, might hold. However, this is an extremely difficult mathematical problem, and as of this writing we do not know the answer (except in very simple cases).

Negative Energy and Classical Fields

In this chapter we have primarily been discussing negative energy in the context of quantum fields. All *observed classical* (nonquantum) fields obey the energy conditions we discussed earlier. However, there are certain *theoretical classical* fields (i.e., they *might* exist) that show up in other areas of physics such as particle physics and cosmology, which violate the weak energy condition. One is called the nonminimally coupled scalar field (NMCSF). That's rather a mouthful! ("Yeah, a friend of mine bought one of those, but it broke the first day he got it.") Rather than go into all the technical details about what that name means, let us just describe how such a field behaves with regard to negative energy, which is, after all, our main purpose here.

Since the field is classical, it is not immediately subject to the quantum in-equalities, which are constraints on *quantum* fields. One might then conceivably use such a field to produce large negative energy fluxes and densities to produce big effects. In the late 1990s, Carlos Barcelo and Matt Visser (then both at Washington University in St. Louis) showed that such a field could in principle violate all the energy conditions. They showed how to construct wormholes using a NMCSF as a source of the negative energy required to hold the wormhole open. Unfortunately, the required parameters that describe the field necessary to achieve this seem to be *enormously*, some would say unphysically, large.[6] Another rather disturbing feature of their wormholes is that the *sign* of Newton's gravitational constant could be different in the two external regions of space connected by the wormhole. (Barcelo and Visser suggested that one might patch up this latter problem by adding some normal matter to the mix.) It is also possible, although not proven, that the field parameters that they need to assume would lead to violations of the second law of thermodynamics as applied to black holes.

However, work by Fewster and Lutz Osterbrink (then also at the University of York) showed that in flat spacetime (and in certain classes of curved spacetime), the NMCSF does obey the *averaged* weak and null energy conditions. Furthermore, they showed that the negative energy associated with this classical field exhibits quantum interest–like phenomena, for example, a negative energy pulse must be overcompensated by a positive energy pulse. Their results also imply that large, long-lasting negative energy is only achievable with uncharacteristically (and very possibly, unrealistically) large fields. This conclusion is consistent with the required field parameters that Barcelo and Visser had to assume in the construction of their wormhole models.

Even though this is a section on classical fields, we close by mentioning some recent results on *quantum* NMCSFs. In a second paper, Fewster and Osterbrink showed that these quantum fields *do not* obey the quantum inequalities, at least in their usual form. If you go back and look at the expression for the quantum inequality bound given earlier in the chapter, you notice that the right-hand side does not depend on the quantum state of the field. Fewster and Osterbrink demonstrated that quantum NMCSFs obey a *weaker* type of quantum inequality in which the right-hand side *does depend* on the quantum state. The energy density can be made arbitrarily negative over an arbitrarily large spacetime region, but only at the cost of an even larger total amount of positive

6. More specifically, the field must take on *trans-Planckian* (i.e., beyond the conjectured scale of quantum gravity) values.

energy. Their results indicate that it is still harder to create negative energy than positive, and that these difficulties increase the larger the energies involved. As of this writing, these results are fairly new, and their consequences have yet to be fully understood.

We have introduced a lot of new ideas in this chapter. Let us try to summarize, as best we can, the current playing field. The averaged weak energy condition holds in flat spacetime when there are no boundaries. The averaged null energy condition holds in flat spacetimes, and has been recently shown to hold even in flat spacetimes with boundaries. However, it fails in some curved spacetimes. On the other hand, it has been conjectured that a suitably modified (i.e., "self-consistent") version of the averaged null energy condition might hold in these cases as well.

As we have discussed, quantum inequalities have been proven for several known free (noninteracting) fields in flat spacetime, such as the electromagnetic field and the Dirac (i.e., electron) field, and for some fields that may exist. Recently, some similar bounds have been proven in curved spacetime. The quantum inequalities, in their original form, do not hold for the Casimir effect or for the interacting fields example of Olum and Graham. However, in the Casimir case (and probably in the Olum-Graham case as well), it is possible to define difference inequalities. These are bounds on the difference in energy in the lowest energy state of the system (its "ground state") and in an arbitrary quantum state. A number of quantum difference–type inequalities have also been proven in curved spacetime. These difference inequalities say that one cannot arbitrarily depress the negative energy below the negative energy ground (i.e., lowest) state of the system.

The Olum-Graham results indicate that the original form of the quantum inequalities do not hold for at least some interacting fields. The analysis of interacting fields is mathematically a much more difficult problem than for free fields. As a result, the state of our knowledge of quantum inequalities with regard to interacting fields in general is still in its infancy, and is a current area of research. For the NMCSF, which might exist in nature, the usual quantum inequalities do not hold, but there are others that do. These other quantum inequality bounds depend on energy scale and become stricter at higher energies.

In the following chapter, we will discuss the implications of the quantum inequalities for wormholes and warp drives.

12

"To Boldly Go . . ."?

Captain, I can't change the laws of physics.

SCOTTY, *Star Trek*

In this chapter, we shall examine the viability of the various spacetime shortcuts discussed in chapter 8. Do the laws of physics limit their behavior or prevent their creation? In appendix 6, we also discuss a famous theorem of Hawking that proves, given some very reasonable assumptions, that negative energy is *always* required to build a time machine in a *finite* region of spacetime.

Curved versus Flat

From our discussion in the last chapter, it would seem that the most likely candidate for the negative energy required for wormholes and warp drives is that associated with certain states of quantum fields. For such states, the quantum inequalities place very strong constraints on the possible configurations of wormhole and warp drive geometries. The quantum inequalities for flat spacetime can be used in curved spacetime as well, provided we limit our sampling times to regions of curved spacetime that are small enough to be considered flat over the time of sampling. The curvature of spacetime is described by a mathematical quantity called the "Riemann curvature tensor." (Bernhard Riemann was a famous nineteenth-century mathematician whose work on curved space geometry paved the way for Einstein's theory of general relativity.)

Any smooth curved surface can be considered "locally flat," that is, flat over a small enough region. For example, the curvature of the earth is not immediately noticeable to us because most of the distances we encounter in everyday life are small compared to the earth's radius, its radius of curvature. When we

< 181 >

draw a triangle on the floor of our laboratory, the sum of the angles add up to 180°, in accordance with familiar Euclidean geometry. That's because our laboratory is small compared to the earth's radius of curvature. By contrast, if we drew a very large triangle on the surface of the earth, with two legs made up of portions of longitude lines, and the third being a portion of the equator, then we would find that the sum of the angles of this triangle add up to more than 180°. (This feature of triangles is one of the properties of a spherical surface that distinguish it from a flat surface.) For a general sphere, the smaller the sphere, the more curved it is. *The smaller the radius of curvature, the "more curved" the surface, and the smaller is the size of a region that can be considered flat. For a region to be small enough to be considered flat, its size in every direction must be much smaller than the radius of curvature of the sphere.*

In the case of four-dimensional curved spacetime, the laws of special relativity hold over regions that are small enough, in space and time, to be considered flat over the duration of any measurements we make.[1] Unlike a sphere, which is described by one radius of curvature, a general curved spacetime can have a number of different radii of curvature, because there can be curvature in more spatial dimensions, as well as in the time dimension. These radii of curvature can be determined from the Riemann curvature tensor for the spacetime.

The advantage of the quantum inequalities is that they contain a sampling function, whose sampling time we can set to be anything we like. Even though the average we take is technically over the entire worldline of an inertial ("geodesic," or, freely falling, in curved spacetime) observer, the only region that really contributes significantly to the average is the part of the observer's worldline that is within the sampling time we choose. Thus, we can apply the flat spacetime quantum inequalities in a curved spacetime provided we *take the sampling time to be much smaller than the smallest radius of curvature of the spacetime.* (Here, we measure the sampling time in ct units, so that it has units of length.) Over that sampled region, the spacetime can be considered flat and the laws of special relativity must apply. The same reasoning holds if the spacetime contains boundaries, for example, mirrored plates (as in the case of the Casimir effect). Consider a (very tiny) observer who is between a pair of Casimir plates. If we choose the observer's sampling time (again, in ct units) to be small compared to the distance to the plates, as measured in his rest frame, then the original quantum inequalities also apply to the Casimir effect. If you were to

---

1. That is, the region should be small enough for tidal forces to be negligible over the time of the measurement.

replace the phrase "distance to a boundary" with "spatial extent of the negative energy density," then the same is likely also to apply to the Olum-Graham example referred to in the last chapter.

Wormholes and Quantum Inequalities

In 1996, Ford and Tom applied the flat spacetime quantum inequalities, using the method outlined above, to Morris-Thorne wormhole spacetimes. By choosing the sampling time to be small compared to the smallest radius of curvature or the distance to any boundaries, they were able to prove very strong constraints on possible wormhole geometries. If the wormhole is macroscopic (e.g., large enough for a human being or a spaceship to pass through), there must be huge discrepancies in the length scales that describe the wormhole. Otherwise the wormhole can be no larger than approximately what is called the "Planck length," which is about 10^{-33} centimeters. The Planck length characterizes the scale below which the presently unknown laws of quantum gravity become important and the predictions of our present theories become unreliable.

By "length scales that describe the wormhole," we mean things like the radius of the throat of the wormhole and the thickness (in the radial direction) of the negative energy region near the throat. For typical cases, Ford and Tom found that for macroscopic wormholes (i.e., large throat size), the negative energy had to be confined to an incredibly thin band around the throat. For example, one type of wormhole postulated by Morris and Thorne was dubbed the "absurdly benign" wormhole. This is because the wormhole geometry was tailored so that an infalling observer would experience no tidal forces. When Ford and Tom applied the quantum inequalities to this wormhole, they found that for a wormhole with a throat radius of one meter (just about large enough for a human being to crawl through), the negative energy had to be concentrated in a band around the throat that could be *no thicker than* about one-millionth the size of a proton! (The size of a proton is about 10^{-13} centimeters, or $\frac{1}{100,000}$ the size of an atom.) Matt Visser previously estimated that the amount of exotic matter required to hold open a meter-sized wormhole is about the equivalent of the mass of the planet Jupiter, but negative in sign (i.e., "minus" the mass of Jupiter). In terms of energy, using $E = mc^2$, this is equivalent to about the total amount of energy produced by ten billion stars in one year, but negative in sign. So it appears that to maintain a wormhole you could just crawl through, you would need minus the mass of Jupiter, confined to a region no thicker than a millionth of a proton radius. The situation does not improve very much if we consider larger wormholes of this type. A wormhole with a throat radius of one light-year would still be required to have its negative energy confined to a region whose thickness is less than a proton radius.

In other research conducted around the same time, Brett Taylor, Bill Hiscock (both then at Montana State University), and Paul Anderson (at Wake Forest University) analyzed the matter/energy profiles for several quantized fields to see if they would be compatible sources for supporting traversable wormholes. In all the wormhole models that they examined, they found that the matter and energy associated with these fields did not have the properties required for maintenance of the wormhole geometry.

Recall that in chapter 9, we mentioned Visser's "cubical" wormholes, which had the negative energy confined to the edges of a cube so that an observer could pass through the wormhole without encountering the exotic matter. If the edges of the cube were extremely thin, such a configuration might satisfy the quantum inequality bounds. One type of object predicted by various theories of particle physics and cosmology, and which may well exist in the real universe, is known as a "cosmic string." This is an immensely dense, but incredibly thin, object of great length. A cosmic string is so dense that a few kilometers of it would weigh as much as several times the mass of the earth. Although cosmic strings have negative pressure, they obey the weak energy condition and don't have negative energy density. Thus they are *not* exotic matter in the sense used here.[2] To support a Visser cubical wormhole, one would need something like a negative energy density cosmic string. However, all the theories known that predict cosmic strings allow only strings with *positive* energy density.

Another issue we have not yet dealt with is how one acquires a wormhole in the first place. Starting with flat spacetime, one would have to "punch a hole" in it to make a wormhole. Nobody has the remotest idea how to do that, or if it is even possible. The theories of quantum gravity currently on the market all suggest some sort of "granularity" of space at the smallest levels (i.e., on the size of the Planck length). The late physicist John Wheeler suggested that one possibility is that space on these scales might be analogous to the foam on an ocean wave. Seen from high above, in an airplane, the ocean surface looks smooth and serene. If we look on much smaller scales, for example, the scale of a single ocean wave seen just above a wave crest, we see the much more

---

2. It should be pointed out that the contrary, *incorrect* statement is made in the popular book *Black Holes, Wormholes, and Time Machines*, by Jim Al-Khalili (London: Institute of Physics Publishing, 1999), 214, 227.

complicated substructure of bubbles, froth, and foam. Wheeler dubbed this picture of space on the smallest scales "spacetime foam." Morris and Thorne suggested that perhaps an arbitrarily advanced civilization might be able to pull a submicroscopic wormhole out of the spacetime foam and enlarge it to traversable size. Of course, no one has any idea how to do that, either. Some years ago, Tom toyed with the idea of whether the rapid expansion of the early universe, "inflation," might naturally enlarge such a tiny wormhole to macroscopic size. The conclusion was that, for a variety of reasons, this did not seem to be a very plausible mechanism.

In 2003, Matt Visser, Sayan Kar (at the Indian Institute of Technology), and Naresh Dadhich (at the Inter-University Centre for Astronomy and Astrophysics in India) suggested that it might be possible to make wormholes with arbitrarily small amounts of exotic matter. To reach their conclusion, they did not assume anything about the source of the exotic matter. If one assumes that the source is the negative energy associated with a *quantum* field, then one can apply the quantum inequalities to these wormholes as well. This was subsequently done by Fewster and Tom, using a somewhat more powerful form of quantum inequality. They found that the Visser-Kar-Dahich (VKD) wormholes also run afoul of the quantum inequalities, and that VKD wormholes of macroscopic size are either ruled out or are severely constrained. This is similar to the earlier conclusion of Ford and Tom. Alternatively, one could assume that the source of the exotic matter is a classical nonminimally coupled scalar field, as in the earlier Barcelo-Visser wormholes, but we saw that those wormholes required enormously large values of the field parameters.

Warp Drives and Quantum Inequalities

After the first application of quantum inequalities to wormholes, Mitch Pfenning and Larry Ford performed a similar analysis for the Alcubierre warp drive spacetime, assuming a quantum field source for the required negative energy. They found that the constraints on warp bubbles are even more stringent than the ones for wormholes. It turns out that the thickness of the bubble wall is restricted by the relation

$$\text{wall thickness} \leq 10^2 \, \frac{v_b}{c} L_{\text{Planck}},$$

where v_b is the speed of the warp bubble, c is the speed of light, and $L_{\text{planck}} = 10^{-33}$ centimeters is the Planck length. So unless the speed of the bubble is *enormously* larger than the speed of light, the bubble wall thickness cannot be much larger than the Planck length. In considering the total amount of negative energy required, Pfenning and Ford calculated that, for a bubble radius of about 100 meters (large enough to fit a spaceship in), the total magnitude of negative energy, $|E|$, required was given by

$$|E| \geq 3 \times 10^{20} M_{galaxy} \, v_b,$$

where M_{galaxy} is the mass of our entire galaxy.[3] So for a 100-meter warp bubble, you need (minus) the mass of about 10^{20} galaxies, which is about 10 powers of 10 (i.e., 10 orders of magnitude) larger than the total mass of the entire visible universe!

Allen and Tom did a similar analysis for the Krasnikov tube and found that the situation was even worse there. We found that to make even a laboratory-size Krasnikov tube (1 meter long and 1 meter wide) would require an amount of negative energy with a magnitude of about 10^{16} galaxy masses. To build a tube that stretched from here to the nearest star would require about (minus) 10^{32} galaxy masses. The constraints on the maximum thickness of the tube walls are comparable to that found for warp bubbles.

Van Den Broeck's "Ship in a Bottle"

Chris Van Den Broeck (then at the Katholieke Universitat Leuven in Belgium) proposed an ingenious idea to dramatically reduce the amount of negative energy required for a warp bubble, to (only!) a few times the mass of the sun. We call this the "ship in a bottle" approach. We can visualize this in the following type of rubber sheet diagram. Imagine an inflated balloon, seamlessly attached to a flat sheet by a narrow tube or neck. The outer radius of the neck represents the radius of the warp bubble. Draw a spaceship on the bottom of the (two-dimensional) surface of the inflated balloon. (Remember that in these diagrams, the rubber sheet *itself* represents space. The regions inside and outside the sheet have no physical meaning, and are only there to allow us to visualize the curvature of the sheet.) In Van Den Broeck's model, the outer radius of the warp bubble is only about 3×10^{-15} meters, about the size of a proton. However, as one proceeds into the bubble, the interior of the bubble opens out into a large macroscopic "pocket" of curved space with a flat region near the center, where the spaceship is placed. This region is connected to the warp bubble by a narrow throat.

---

3. The mass of our galaxy is roughly about a hundred billion times the mass of the sun.

Van Den Broeck's modified geometry for the warp drive reduces the total negative energy requirements down to a few times the mass of the sun, compared to 10^{20} galaxy masses. However, the problem of large negative energy densities, implied by the quantum inequalities, remains. That is, the thickness of the warp bubble walls is limited by the same restrictions as in Alcubierre's original model to be no thicker than a few Planck lengths. Also, it is unclear how one could warp space enough to put the ship in the "bottle" in the first place—let alone how to "unwarp" space to get it out! And one still has the problem, present in the original model, of the inability to steer the bubble from the inside. Pierre Gravel and Jean-Luc Plante, at the Collège Militaire Royal du Canada, subsequently performed an analogous modification of the Krasnikov tube, modeled after Van Den Broeck's paper on the Alcubierre warp drive, with similar results.

More Trouble for Warp Drives

Francisco Lobo, of the Centro de Astonomia e Astrofísica da Universidade de Lisboa in Portugal, and Matt Visser performed a detailed analysis of the Alcubierre and Natário warp drives. They did not assume anything about the source of the exotic matter, that is, whether it was quantum or classical in nature, so their results are quite general. Lobo and Visser pointed out that the weak energy condition violation persists for *arbitrarily low* bubble speeds. This means that the need for exotic matter is not just associated with superluminal velocities. It seems to be related to the fact that these warp drive mechanisms are examples of "reactionless drives," which appear in science fiction.

What does this mean? All propulsion systems, such as conventional rockets, work on the principle of Newton's third law of action and reaction (which is the law of conservation of momentum). This law states that for every action force there is an equal and opposite reaction force. (The action and reaction forces act on *different* bodies, which is why they don't cancel one another out.) A rocket works by expelling material (matter or radiation) out the back. The rocket exerts a force on the fuel by hurling it backward, and the fuel exerts an equal but opposite force on the rocket, which drives the rocket forward.[4] The warp drives do not work this way; there is no ejecta thrown backward to propel

---

4. A common misconception, easily arrived at by watching a space launch, is that a rocket works by "pushing" against the surface of the earth. If that were true, a rocket could not function in empty space, because there would be nothing to "push against."

the spaceship forward. The spaceship sits at rest inside the bubble and the bubble simply carries the spaceship along with it. This would also be true for our example of the two-particle negative- and positive-mass system that self-accelerates (discussed in chapter 11). In that case also there is no fuel ejecta that produces a reaction force on the system driving it forward. Looking at it in this way, it is perhaps not too surprising that the warp bubbles require negative energy for even arbitrarily small speeds.

Lobo and Visser also showed that the total negative energy of the warp field must be an appreciable fraction of the positive mass of a spaceship placed within the bubble. In order for the total negative energy in the warp field not to exceed the mass of the spaceship, they found that the bubble speed had to be extremely low. It is worth mentioning again that Lobo and Visser's conclusions are independent of any assumptions using quantum inequalities, since they did not assume anything about the nature of the negative energy.

Ways Out?

Let us return to the subject of the quantum inequality bounds on wormholes and warp drives. Those of you who are uncomfortable with the conclusions we have drawn might be wondering if there are any ways to circumvent them. We will discuss a few here. One possibility would be to try to superpose (add the effects of) many different fields involving negative energy. While each field individually might obey a quantum inequality bound, by putting many such fields together it might be possible to overwhelm the bound. However, one can estimate how many fundamental fields in nature would be required to overturn, say, the restrictions on traversable wormholes. For example, a calculation shows that for a 1-meter wormhole, one would need 10^{62} fundamental fields! (If you are a string theorist, and you believe that there really are 10^{62} or more fundamental fields in nature, then shame on you!)

Other possibilities might include classical fields with large negative energies, such as the nonminimally coupled scalar field. However, as we saw in the last chapter, attempts to make traversable wormholes from them involved enormously large values of the field parameters. As Fewster and Osterbrink recently showed, the quantized version of this field does not obey the usual quantum inequalities (although it does obey a weaker form, the implications of which are currently under study).

Recall also that the quantum inequalities were proven for *free* fields. From the work of Olum and Graham, we learned that *interacting* fields are unlikely to obey the usual quantum inequalities. It is possible that they obey some other form of quantum inequality, as in the case of the quantized NMCSF. Calculations involving interacting fields are mathematically much more complex than those for free fields, so the situation is unclear at present, although some people are currently investigating this problem.

Lastly, we mention the possibility that the "dark energy," which is driving the recently discovered accelerated expansion of the universe, might violate the weak energy condition. For many years, astronomers had expected that the gravitational attraction of all the galaxies on one another would gradually slow down the expansion of the universe over time. However, observations of distant galaxies made in the late 1990s indicated, to most people's great surprise, that the expansion of the universe was speeding up!

The question is: What is causing this? Since we don't know, we simply call it "dark energy," because its only manifestation seems to be gravitational. Whatever it is, it must have a repulsive gravitational effect, which can be modeled in several ways. One of the most popular is a "cosmological constant," an extra (but mathematically allowed) term in Einstein's field equations of gravitation. Originally introduced by Einstein to keep the universe static, as it appeared to be at that time, it was abandoned soon after Edwin Hubble discovered in 1927 that the universe was expanding. Einstein called it "the biggest blunder" he ever made. In light of the present situation, perhaps he was wrong about that.

The cosmological constant term acts as a repulsive force at large distances. It has negative pressure, but *positive* energy density. So it is not exotic matter, in the sense that we have been using the term. (Not to say that it isn't weird!) Other proposed models for dark energy include new kinds of fields, possibly exotic, with negative energy density. At current writing, these weak energy condition-violating fields are not inconsistent with the observational data. At present, the riddle of dark energy is one of the most intensely investigated problems in cosmology, and the answer is far from clear.

If we had to bet on any of these possibilities for avoiding the conclusions of the quantum inequalities, I think we might pick either interacting fields or some form of exotic dark energy. But at this point our bet would be a small one.

Time Machine Destruction and Chronology Protection

In the early 1990s, Kip Thorne was visiting the University of Chicago to give a talk on his work on wormhole time machines. It was pointed out by colleagues

Bob Geroch and Bob Wald that once the time travel horizon forms, a beam of radiation could circle through the wormhole over and over again, an arbitrary number of times, piling up on itself, until the huge energy density thus produced would destroy the wormhole. This worried Thorne for a while until he realized that, upon each traversal of the wormhole, the circulating radiation beam would get defocused due to the diverging effect of the wormhole on light rays (recall that this effect was discussed in chapter 9). As a result, he found that this defocusing property would dilute the energy in the beam more and more on each pass so as to overwhelm the effect of the pileup. The wormhole time machine was safe. Or was it?

Sun-Won Kim, from Ewha Womans University in Korea, and Thorne published an article in 1992 with a surprising conclusion: although Thorne had earlier concluded that a beam of classical radiation circulating through the wormhole would not lead to its destruction, he and Kim now turned their attention to the effects of *vacuum fluctuations* circulating through the wormhole. They found something totally unexpected. Unlike classical radiation, the vacuum fluctuations were *not* defocused by the wormhole. Kim and Thorne's calculations showed that the vacuum fluctuations would travel through the wormhole over and over again, piling up on themselves and causing the destruction of the wormhole. And vacuum fluctuations are everywhere; they can't be "turned off." So it appeared that the wormhole time machine would self-destruct as soon as it was formed.

However, there also seemed to be a potential loophole. The energy density pileup of vacuum fluctuations becomes infinitely large for only an infinitesimal instant of time and then dies down again. It peaks right when the wormhole first becomes a time machine. Kim and Thorne were using the techniques of quantum field theory in curved spacetime to analyze this problem. This is also called the theory of semiclassical gravity. It treats matter and energy according to the laws of quantum mechanics, but treats gravity according to classical (nonquantum) general relativity. Although we know that this theory is valid over a wide range of circumstances, it cannot be a complete theory. Ultimately, we expect it to be superseded by a quantum theory of gravity. The catch is that the laws of quantum gravity appear to kick in right when the energy density of the vacuum fluctuations becomes large enough to destroy the time machine. The key question is: Will those laws cut off the growth of the energy density buildup of the vacuum fluctuations *before* it becomes large enough to destroy the time machine? After much back-and-forth discussion between Thorne and

Stephen Hawking, it was concluded that the answer could well be "no." Unfortunately, we cannot know for sure until we have the laws of quantum gravity in our possession. So at present the answer is not completely clear-cut.[5]

Hawking subsequently proposed his "chronology protection conjecture": *the laws of physics prevent the formation of time machines for backward time travel.* He initially thought that the mechanism for this would be the energy density pileup of circulating vacuum fluctuations on the time travel horizon, as Kim and Thorne had found. Hawking's proposal was that this process would be the chronology enforcer for *any* kind of time machine, wormhole or otherwise. However, since his initial proposal, counterexamples have been found. That is, there are particular model spacetimes one can cook up in which a time travel horizon forms, but the energy density of the vacuum fluctuations on the horizon does *not* blow up as one approaches the time travel horizon. Therefore, if nature does protect chronology, it cannot always be by this mechanism.

In 1997, Bernard Kay, Marek Radzikowski, (then both at the University of York), and Bob Wald gave a very strong mathematical argument in favor of Hawking's chronology protection conjecture, using the techniques of quantum field theory in curved space, that is, semiclassical gravity. The Kay-Radzikowski-Wald results showed that the quantity that represents the matter/energy (and other stuff, like pressures and fluxes) of a quantum field, the so-called quantum stress-energy tensor,[6] either blows up (as in the Kim-Thorne wormhole time machine) or is undefined, on a time travel horizon, for any physically sensible quantum state of the field.[7] If it blows up, we would expect that the back-reaction on the spacetime would destroy the time machine. If it is undefined, then that says that a sensible quantum field theory description on the time travel horizon is unattainable. Presumably, one would then need the laws of quantum gravity to determine what happens. The Kay-Radzikowski-Wald results would seem to be evidence in favor of chronology protection.

However, Visser pointed out that to determine when the laws of quantum gravity kick in depends not just on when the quantum stress-energy tensor blows up. It also depends on when quantum fluctuations of spacetime,

---

5. More detail on this topic can be found in Kip Thorne, *Black Holes and Time Warps: Einstein's Outrageous Legacy* (New York: W. W. Norton, 1994), chapter 14.

6. *Technical note:* Here, we really mean, more precisely, the "expectation value" of the stress-energy tensor.

7. The previously mentioned counterexamples have the peculiar property that the quantum stress-energy tensor is well behaved up to, but not right *on,* the time travel horizon.

predicted by the uncertainty principle, become very large. He pointed out that these two regimes need not be the same.

Visser defined what he called the "reliability horizon," up to which we can trust the laws of semiclassical gravity, but past which one would need the laws of quantum gravity to decide what's going on. He argued that if the time travel horizon lies outside the reliability horizon, then the Kay-Radzikowski-Wald results of semiclassical gravity are trustworthy. However, if the time travel horizon lies *inside* the reliability horizon, then the laws of semiclassical gravity break down *before* one gets to the time travel horizon. In that case, the laws on which the Kay-Radzikowski-Wald calculation is based have already broken down before you even reach the time travel horizon, so what the Kay-Radzikowski-Wald results tell you about what happens there is suspect. If this is the situation, then we need the full laws of quantum gravity to resolve the issue.

For example, if in your time machine spacetime, the quantum stress-energy tensor starts to increase rapidly as you approach the time travel horizon, but *has not yet* blown up when you have reached the *reliability* horizon, then you can't determine what happens past that point without the laws of quantum gravity. Maybe quantum gravity cuts off the explosion and saves the day, or maybe it doesn't, and the time machine is destroyed. Without those laws, you can't know for sure. Visser then went on to give a convincing argument that, in fact, the time travel horizon generally lies *inside* the reliability horizon (see the cartoon in figure 12.1). So it appears that to settle the question of whether the universe protects chronology, and so outlaws time machines, will require knowledge of the laws of quantum gravity. Visser is careful to emphasize that his result does not imply that one could actually succeed in building a time machine. Rather, what it implies is that our current knowledge is insufficient to *definitively* answer the question. On the other hand, the Kay-Radzikowski-Wald result implies that a time machine spacetime cannot be described semi-classically. It is rather hard to imagine what it would be like, or even mean, to "travel" through a region of spacetime that could *only* be described by the laws of quantum gravity. Of course, to have the paradoxes of time travel, one would only need to be able to send signals through such a region, not necessarily to travel through it oneself. But even that may be problematic in a regime where space and time do not have their usual forms.

One of the examples in which the "back-reaction" of the vacuum fluctuations on the spacetime geometry can be made as small as possible is Visser's ingenious "ring of wormholes" spacetime. Suppose that we have a number of

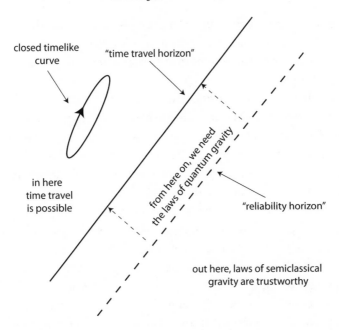

FIG. 12.1. Time travel versus "reliability" horizons. Beyond the reliability horizon, spacetime cannot be treated semiclassically and we need the laws of quantum gravity to determine what actually happens.

wormholes, N, arranged in a ring. Visser sets up the parameters of his model so that each *individual* wormhole is not itself a time machine. An observer passes through one wormhole and then journeys through normal space to the mouth of the next successive wormhole in the chain, and so on. Visser then goes on to demonstrate that, by starting with a system of such wormholes, no subgroup of which is a time machine, one can turn it into a time machine. Furthermore, for this ring of wormholes, his detailed calculations show that the back-reaction of the quantum fluctuations can be made as small as one likes, right up to the reliability horizon, by letting the number of wormholes, N, become arbitrarily large. The more wormholes one adds, the smaller the back-reaction. Visser's conclusion is not that he has succeeded in building a time machine, but that the laws of quantum gravity will be needed to decide the issue. This is because of the problem of reliability versus time travel horizons discussed earlier.

Hawking has retreated somewhat from his original suggestion that the energy density of vacuum fluctuations will always be the mechanism that protects chronology. He has since suggested other possible ways that time machines might be forbidden, which are a bit too technical to be described here. As of this writing, Hawking still appears to feel that *some* form of chronology protection is likely to be true.

If so, perhaps the present situation is analogous to attempts to build perpetual motion machines prior to the discovery of the laws of thermodynamics. Careful analyses of such machines show that they always failed for one reason or another, but the *particular* reason varied from machine to machine. We now know that the underlying principles that they all violate are the first or second laws of thermodynamics. Maybe there is a similar principle at work here that always protects chronology, but not necessarily always by the same mechanism. However, for now at least, Hawking's proposal still remains a conjecture—albeit a reasonable one, in our opinion.

One of the themes of this book has been the connection, due to special relativity, between the possibility of superluminal travel and that of backward time travel. With this connection in mind, we might ask whether the chronology protection conjecture, if true, also necessarily forbids superluminal space travel. Suppose you can create an object—a wormhole, warp bubble, whatever—that allows superluminal travel. Then, as discussed in chapters 6 and 9, the principles of relativity ensure that you can create an arrangement of two such objects operating in opposite directions, one at rest in each of two different inertial frames, which will have the following property. An object or person following a worldline through the two objects in succession will return to the starting point in spacetime. That is, the worldline would be a closed timelike curve, the formation of which is precisely what is forbidden by the chronology protection conjecture.

However, even if true, the conjecture does not imply that you can't make wormholes. Rather, it implies is that you can't have a configuration of wormholes (or warp bubbles or Krasnikov tubes) that would lead to the formation of a closed timelike curve. As long as the configuration is such that a would-be time traveler will return to her starting point *after* she left, her worldline will not be a *closed* timelike curve. Superluminal travel, however, can still occur. Our point is that, while faster-than-light travel *can* be arranged so that it is possible for a traveler to return to her starting event, it *need not be* arranged that way. A chronology protection mechanism would rule out the former possibility but not the latter. But, as we have seen, you do have to be careful in choosing the routes of your wormholes and warp bubbles.

In the future, the human race may well want, or need, to expand beyond the solar system. A world with a warp drive, but without backward time travel, and its associated paradoxes, might well represent a best-case scenario. Unfortunately, even though chronology protection does not appear to forbid warp drives, this does not eliminate the discouraging prospects for the existence of wormholes or warp bubbles suggested by the quantum inequalities that we have discussed.

13

Cylinders and Strings

To be accepted, new ideas must survive the most
rigorous standards of evidence and scrutiny.

CARL SAGAN, *Cosmos*

Rolled-Up Universes

A very simple example of a universe that
has closed timelike curves is simply flat
spacetime with the time dimension wrapped into a circle. We can make a two-
dimensional model of such a universe as follows. Take a piece of paper (which
we can think of as representing a piece of infinite flat spacetime) and roll it into
a cylinder. Let the time axis be a circle *that wraps around the cylinder.* (Note that
although our cylinder is three-dimensional, the surface of the cylinder, which
represents the universe in this model, is two-dimensional.)

An observer who has the time axis as his worldline returns to the same
moment in space and time. Lines that run along the cylinder, parallel to its
axis, represent spacelike surfaces (in this two-dimensional model). A moving
observer's worldline will wrap around the cylinder at an angle of less than 45°
with respect to the time axis, just as in flat spacetime. In fact, this cylindrical spacetime is flat, because the geometry of a cylinder is exactly the same
as that of the flat piece of paper. To see this, note that the sum of the angles
of a triangle drawn on the flat piece of paper remains equal to 180° even after
the paper has been rolled up into a tube. Recall that if the space was curved,
the sum of the angles of a triangle would sum to either more than or less
than 180°.

What has been changed is the "topology" of the piece of paper, that is,
loosely, how different points on the cylinder may be connected with one another. Note that on a flat piece of paper, it is possible to take *any* circle that
you draw on it and contract it continuously into a point while remaining on

< 196 >

the paper's surface. However, on a cylinder, a circle that wraps around the symmetry axis of the cylinder cannot be contracted into a point while remaining on the surface of the cylinder. So we can always create examples of spacetimes with closed timelike curves simply by artificially fiddling with the topology of spacetime. This is not to say that such models should necessarily be taken as serious models of our universe; we have no reason to believe that our universe is cylindrical. Such models merely provide more examples for physicists and mathematicians to examine the consequences of Einstein's equations.

Notice that, in practice, our cylindrical universe above cannot be created starting from a finite region of spacetime. Either our universe is "born" with that structure or it isn't. This is a characteristic of all of the spacetimes we discuss in this chapter that contain closed timelike curves.

Our cylindrical universe has some interesting properties. Keep in mind that, in actuality, the cylinder has no "ends." The model illustrates only a finite portion of an infinite cylinder. Consider the worldline of a moving observer that winds around the cylinder. We can ask: on a given spacelike slice, is there one observer or many? On an infinite spacelike surface, there will be many (in fact, an infinite number of) copies of such an observer at a given instant of "time," that is, on a single spacelike slice. There will be one copy located at every point where the worldline intersects the spacelike surface. This observer returns to the same point in time, but at different positions in space. On the other hand, each of these copies of the observer will be a different age, according to the observer's own proper time. So the question of whether, on a given spacelike slice, there are many observers or only one rather depends on exactly what you mean by the question.

Since the cylinder is infinite, we can talk about the number of copies of the observer per unit length (in our two-dimensional model) on a spacelike surface. For a given length of cylinder, the number of copies depends on the velocity of the observer. Curiously, there are *more* copies per unit length the *slower* the observer moves. The *minimum* number of copies per unit length is determined by the number of intersections that a light ray, oriented at 45° to the time axis, makes with the spacelike surface. We can consider this to be the limiting case of an observer who moves arbitrarily close to the speed of light. One rather strange feature of this spacetime is that the slower you go the more copies of you there are per unit length. However, in the other limiting case where your velocity goes to zero, there is only one copy of you on the spacelike slice! This is because the time axis intersects the spacelike surface at only one point.

A "Rotating" Universe

A more sophisticated example is due to a very famous German mathematician named Kurt Gödel. In 1949, Gödel considered an infinite universe made of rotating dust.[1] He discovered that in such a universe, any circle of sufficiently large radius would be a closed timelike curve. How large the radius had to be was determined by how fast the universe was rotating. Such a universe would certainly be an interesting place to live, and the equations of general relativity seem to make it clear that such a universe could exist without contradicting any of the laws of physics, as we know them. However, and probably fortunately for us, that is not the universe we live in. Despite having closed timelike curves, there is no exotic matter in Gödel's universe. However, it does not violate Hawking's theorem, because it is infinite in size and thus (obviously) cannot be constructed in a finite time. The universe would have to have this structure *ab initio*. Observations of distant galaxies strongly indicate that the universe, in fact, is *not* rotating in the way envisioned by Gödel.

Cylinder Time Machines

In the rest of this chapter, we will consider time machines that involve the presence of one of several kinds of infinitely long string-like or cylindrical systems containing rotating matter or energy. (Here, we mean infinitely long cylinders existing in space, as opposed to the infinite cylindrical-type *universes* considered at the beginning of the chapter.) We know more about infinitely long cylinders than those of finite length, since it is easier to solve the difficult Einstein equations in the infinite length case (where you don't have to worry about the ends of the cylinder). In that case the solution does not depend on z, the axis of symmetry of the cylinder; wherever you are along the z direction, you still see the cylinder stretching off to infinity in both the positive and the negative z directions. Another way of saying it is that no matter where you are along the axis of the cylinder, the spacetime surrounding the cylinder looks the same.

In some cases it is possible, by running in the proper direction (say, clockwise) around a circular path enclosing the infinite cylinder in question, to

1. "Dust" actually has a technical meaning in general relativity. It refers to a cloud of randomly moving electrically neutral particles whose speeds relative to one another are small compared with the speed of light. The mass-energy of the dust particles is very large, compared with the pressure, which can be taken to be 0.

return to your starting point in space before you left. You may have to run quite fast to do this, but you won't have to exceed the speed of light relative to your immediate surroundings. Thus, you can travel into your own past, even though, throughout the process you will see time "flowing" forward in the usual way in your immediate neighborhood. Having gone around the circle, if you now wait around at the starting point for a while, perhaps sitting on a couch and reading a good science fiction story, you will then find yourself back at your starting point in both space *and time*, able to greet your slightly younger self who has not yet started running. In other words, these infinite cylinders are encircled by closed timelike curves. Therefore, if you found one of these systems you would, in fact, have found a time machine.

None of these systems, however, provide a practical recipe for actually building a time machine, since you can't hope to construct an infinitely long cylinder in a finite amount of time or in a finite region of space. In the theory of electromagnetism, one often studies infinitely long systems because they provide a good approximation to the case of a long finite object, as long as your distance from the object is small, compared to its length. You can get some feeling why this is true if you imagine you put your eye very close, say, an eighth of an inch, to the midpoint of a yard stick. It will look to you like the yard stick runs off forever in both directions, whereas from the other end of a football field the yard stick would appear very short.

Our infinitely long cylindrical time machines have no region of negative energy density. For the case of an infinitely long cylinder, where the time machine cannot be built in a finite region of spacetime, this does not violate Hawking's theorem. However, if we made the time machine just very long, but not infinite, Hawking would tell us that there could be no closed timelike curves, that is, no time machines. An infinitely long cylinder, if it has no negative energy density, is qualitatively different with respect to time travel from one of finite length, however long it might be, because of Hawking's theorem.

Thus, the models we discuss in this chapter do not provide us with any guidance as to how we might build a time machine, though they do provide additional insight into how time machines can occur in the context of general relativity. The last model we look at is interesting because it involves objects, which, while infinitely long and, hence, unbuildable, might very well have been produced in the very early universe, in the first minute fraction of a second following the big bang. But we will first look at examples of time machines involving infinitely long rotating cylinders of matter or energy.

The first example dates all the way back to 1937, when the Dutch-born physicist W. J. van Stockum considered an infinitely long cylindrical column of rotating dust. Van Stockum assumed that the density of the dust column and its speed of rotation were just such that the column was held together by the mutual gravitational attraction of the dust particles without the need of a vessel to contain them. Much later, in the 1970s, it was pointed out by Frank Tipler (then a graduate student at the University of Maryland) that as long as the speed of rotation of the dust column as a whole was high enough there were closed timelike curves surrounding the column at certain distances from its center that could be calculated, given the column's speed of rotation. An observer moving along such a curve at a sufficiently high speed, which could be less than the speed of light, would indeed return to the starting point before he or she had left.

We observe that there is no exotic matter present in van Stockum's example. The energy of the dust column is given by the total mass of the particles through the Einstein relation $E = mc^2$ and by the kinetic energy that they posses by virtue of being in motion. These are both positive, so there is no negative energy density present. As we have already noted, this does not bring down the wrath of Hawking. Due to the infinite length of the column, such a time machine cannot be constructed in a finite region of spacetime; hence, its construction without exotic matter does not violate Hawking's theorem. But for this very reason the van Stockum time machine is largely of only theoretical and mathematical interest. To construct one would take an infinite length of time, making it of somewhat limited practicality.[2]

Mallett's Time Machine

Another more recent example of a rotating cylinder–type time machine is due to Ronald Mallett. He has discussed this in an article published in the journal *Foundations of Physics* in 2003, and also in his autobiographical book *Time Traveler* (2006). Due to the interest this work has generated in the popular press, we will give a somewhat more extended treatment of this particular topic.

Mallett found a solution of the Einstein equations outside an infinite cyl-

---

2. In 1976, Tipler argued, on the basis of the infinite van Stockum cylinder solution, that a sufficiently long *finite* rotating cylinder would provide the basis for a time machine. However, a year later, he proved a general theorem that showed that would *not* in fact be the case. His theorem showed that in order to get a time machine with a finite cylinder, one would need to violate the weak energy condition (i.e., have negative energy) or have a singularity. Tipler's theorem is related to the more powerful one that Hawking proved in 1992, which we discuss in appendix 6.

inder of circulating light, which did indeed contain closed timelike curves. He suggested that a *finite* cylinder of laser light, carried perhaps by a helical configuration of light pipes around the z axis, could be used as the basis of a buildable, working time machine. However, Mallett's model is fundamentally flawed.

The Mallett solution is independent of z, the coordinate along the axis of the cylinder, meaning that it applies in the case of an infinitely long cylinder. As in the previous cases of the van Stockum and Gödel solutions, the Mallett solution contains no region of negative energy density. Hence, it has the same problem as those solutions. On the basis of the Hawking theorem, one would not expect that a *finite*-length cylinder of circulating light could be the basis of a time machine, while a cylinder of infinite length cannot, by definition, be constructed in a finite length of time. Professor Mallett makes no reference to the Hawking theorem in either his paper or his book. In the book, however, he discusses the possible construction of a time machine, using a finite-length cylinder of circulating light, assuming its behavior would approximate that predicted by his solution for an infinite cylinder. He also discusses possible experiments to detect backward time travel produced by such a machine. All of this would be physically relevant only if the Hawking theorem (the validity of which is generally recognized) could be evaded in some way. For the infinite cylinder, Hawking's theorem is evaded because the time travel horizon is not compactly generated (see appendix 6). In the case of a finite cylinder, the only obvious way to evade the theorem would be to include some region of negative energy density. Our experience with studies of other systems and the constraints of the quantum inequalities suggests that one is not likely to be able to do this in an easy or offhand way, if at all. In summary, the Hawking theorem rules out the construction of a time machine using a finite cylinder of laser light, if one accepts the assumptions of the theorem, which we've tried to argue in appendix 6 are reasonable.

In 2004, Ken Olum, together with Allen, wrote a paper that was published in *Foundations of Physics Letters* examining the Mallett model more closely. They confirmed that the model provided a solution to the Einstein equations and that it predicted the existence of closed circular timelike curves encircling the outside of the cylinder. (In Mallett's published solution, there are no closed timelike curves in the space *inside* the cylinder.) A person or object traveling around one of the curves would return to the same starting point in space and time, thus opening the possibility of encountering a younger self. In other words, there were indeed time machines in the model when the cylinder was of infinite length. However, in addition to the problems associated with Hawking's theorem, Olum and Allen found that the model has two additional problems, one practical, and one theoretical.

The practical difficulty arises because the model also predicts the ratio of the radius R of the closed timelike curves to R_0, the radius of the circulating light beam. Taking the results from Mallett's paper, one finds that the numerical predictions of the model itself for this ratio absolutely rule out any possibility that the model can ever lead to the production of a time machine. For certain reasonable assumptions about the laser power and the size of the system, that ratio obeys the inequality

$$R / R_0 > 10^{\left(10^{46}\right)}!!!!$$

In particular, this number assumes a laser power of 1 kilowatt, a light cylinder radius of 0.5 meters, and a radius of the mouth of the light pipe through which the laser beam travels of about 1 millimeter (a schematic diagram is shown in figure 13.1). These are the numbers used by Olum and Allen in their paper analyzing the Mallett model, and we will stick to those numbers. One megawatt = 10^3 kilowatts might be a bit more reflective of the present state of laser technology. However, as we will see, the particular values of these numbers are completely irrelevant.

The number on the right side of the above inequality is unimaginably, incomprehensibly large. We have become used to hearing about trillions since the economic crisis of October 2008 and still think that is a pretty big number. However, a trillion is only 10^{12}. Not only does the number on the right side of our inequality contain 10^{46} rather than 10^{12}, which would be impressive enough, but 10^{46} isn't the number R/R_0 itself—it's just the *exponent*, the number of times 10 must be multiplied by itself to get. In other words, we get $10^{46} = \log(R/R_0)$. You may remember that when we discussed logarithms in chapter 7 in connection with the definition of entropy, we pointed out that if N is a very large number, log N, while it may also be large, is much smaller than N. That is, $\log(R/R_0)$ is much smaller than R/R_0.

Suppose we let R be as big as the radius of the visible universe, which is about 10^{10} light-years, the greatest distance a light signal emitted at the time of the birth of the universe, about 10^{10} years ago, could travel and still reach us today. Now, a light-year is about 10^{16} meters, therefore, the radius of the visible universe is about 10^{26} meters. In addition, suppose we were really clever and built a Mallett time machine whose circulating light beam had a radius equal to

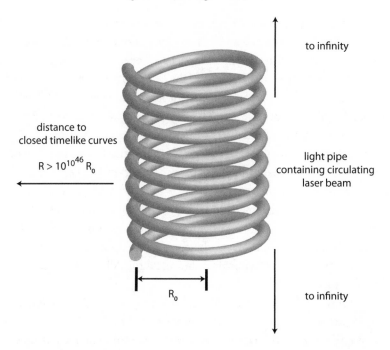

to infinity

distance to
closed timelike curves

$R > 10^{10^{46}} R_0$

light pipe
containing circulating
laser beam

R_0

to infinity

to infinity

FIG. 13.1. Mallett's circulating light cylinder time machine. The closed timelike curves in this model only occur at distances from the cylinder that are unimaginably larger than the size of the visible universe. This problem cannot be fixed by increasing the power of the laser light.

the radius of an atom, about 10^{-10} meters. This situation would give the largest value of R/R_0 we could hope to achieve, even theoretically, for such a the time machine, but it would give an exponent of only 36 rather than 10^{46} in R/R_0 (i.e., it would give $R/R_0 \approx 10^{36}$). This is still a factor of 10^{10}, or ten billion times the size of the visible universe. So the *predicted closed timelike curves, and therefore, the crucial feature of the time machine itself, lie beyond the boundary of the visible universe by a huge factor.* This implies that one cannot build a time machine of this type using even an infinitely long cylinder of circulating laser light.

Just for fun, let's examine how such a humongous number as $10^{\left(10^{46}\right)}$ arises. To lapse into a figure of speech that has entered the lexicon recently, it is the result of a perfect storm of large numbers or small numbers in denominators. For reasons we won't try to go into, the condition for closed timelike curves to arise in the Mallett model, as given in his article, is $K \log R/R_0 = 1$, where K is a dimensionless constant (i.e., one that has no units) of the order of Gm/c^2, where m is the mass per unit length of the laser beam. Here, G is Newton's constant, which appears in his law of gravitation and was introduced in chapter 8, and G/c^2 is of order 10^{-28} meters per kilogram. We can rewrite m as ε/c^2, where ε is the *energy* per unit length in the beam. The quantity we usually know about a laser is P, the power of the laser, which is the energy passing through a given plane perpendicular to the laser beam per second. The energy per unit length turns out to be $\varepsilon = P/c$, and $m = \varepsilon/c^2 = P/c^3$ (these equations are derived in appendix 7, for readers who are interested in the details). Thus, in addition to the small value of G/c^2, we have three extra factors of $1/c$ in the denominator of K. Opposing all these negative powers of 10, there is a puny little factor of 10^3, since we take $P = 1$ kilowatt $= 10^3$ watts, and finally an extra puny little 10^3 derived by a geometrical argument in Olum and Allen's article (this is also derived in appendix 7). This converts m, the mass per unit length along the laser beam as it winds *around* the z axis, to the total mass per unit length *along* the z axis in the circulating laser beam. Putting all this together, one finds $K \approx 10^{-46}$, or $1/K \approx 10^{46}$.

This very small value reflects the combination of two factors. First, gravitational forces are very weak, which means that the value of G/c^2 is very small. Second, the amount of mass in even a very powerful light beam is tiny, compared to the same volume of ordinary matter, because of the small value of P/c^3, even when P is quite large. Given these two factors, and the fact that creating a closed timelike curve involves a rather drastic distortion of spacetime, it was possible to predict intuitively, even without a detailed calculation, that the effect produced by a cylinder of circulating light would likely be very small.

However, as significant as the large value of $1/K$ is, the incredibly large value of R/R_0 could not have been intuitively predicted until one knew the details of the Mallett solution. What moves the result for R from the huge to the humongous is that not only is K very large, but that in Mallett's equations it is not R but the logarithm of R that is proportional to $1/K$. This means that R/R_0 is not just equal to $1/K$, but to the absurdly large value $10^{1/K}$.

It may seem surprising that the effects of such a very small mass of the light beam show up at very large distances rather than very close to the cylinder of light. The reason is that we are dealing with the unphysical situation of an infinitely long cylinder that looks just as long no matter how far you get from it. The fields produced by such objects sometimes actually *increase* very slowly as the distance becomes infinitely large. In fact, this happens in the van Stockum-Tipler dust column model referred to above. In that model, in addition to depending on distance, the gravitational field depends on the rate of rotation.

One finds that there are no closed timelike curves when the dust rotates very slowly. When the frequency of rotation—and therefore the kinetic energy and thus the mass of the source—is increased, closed timelike curves appear, and they are first formed at infinity.

As we have noted, it is common practice in many areas of physics to model finite sources by infinite ones. This is regarded as a good approximation for radial distances much less than the length of the cylinder and away from the ends. However, in general relativity, it is dangerous to extrapolate the behavior of a finite source from an infinite one. This is because, in Einstein's theory, matter and energy curve the very structure of spacetime. An infinite source can curve the *large-scale*, or "global" structure of spacetime in ways that a finite source cannot.

So what about time machines of finite length (if we ignore for the moment that they are forbidden by Hawking's theorem because of the lack of exotic matter)? Such a device is well approximated by one of infinite length only at radial distances R that are much less than L, where L is the length of the cylinder (i.e., at radial distances from which the apparatus "looks" to be essentially infinite in length). Suppose—Hawking's theorem and possible general relativistic complications notwithstanding—that a long, finite-length circulating light beam could in some circumstances be approximated by one of infinite length. Then the prediction of closed timelike curves would apply only to a circulating light cylinder whose length exceeded the predicted radius of the closed timelike curves in Mallett's model. Thus, we would need an apparatus whose length, though finite, would still be greater than $10^{\left(10^{46}\right)} R_0$. Obviously building such an apparatus is not even remotely possible. Even if we relax the infinite length requirement, constructing the "finite" length apparatus needed to create a Mallett time machine would be physically impossible.

In ending this part of the discussion, it is very important to emphasize that the problem is of such gigantic magnitude that *no technological fix is conceivable*. For example, could one just increase the power of the laser? It turns out that changing P from 1 kilowatt to 4×10^{23} kilowatts, which is the total power output of the sun itself, would change our result to $R/R_0 \approx 10^{\left(10^{20}\right)}$, which is again incomprehensibly large, compared to the value of 10^{36} we found for the ratio of the radius of the visible universe to R_0. Even if we increased the laser power to about 10^{35} kilowatts, which is roughly equivalent to the total power output of the 200 billion stars in our galaxy, we would still get $R/R_0 \approx 10^{\left(10^{11}\right)}$!

The preceding discussion, while it eliminates any actual possibility of using a circulating light beam to produce a terrestrial time machine, suggests that, in principle, such a device could produce closed timelike curves, albeit at distances that lack any relevance to observable phenomena. Even that theoretical possibility might be of interest, since we know so little about when, or if, closed timelike curves can be produced.

However, as mentioned earlier, there is another questionable aspect of the Mallett solution. What happens when you turn the laser off by setting P, and hence, K, equal to zero? You would expect to get just flat spacetime at $R > R_0$. For example, this is what happens in the van Stockum solution when the mass/energy of the cylinder goes to zero. Another example is the Schwarzschild solution, describing the spacetime outside of a spherical star, in the case when the mass of the star goes to zero. Again, what you get is flat spacetime. This is the behavior one would expect for a physically realistic source. In the Mallett solution, when you turn off the laser, you get something that looks rather funny. Allen assumed that this was just because Mallett was using a nonconventional coordinate system. Doing this, if you want, is perfectly in accord with the rules of general relativity.

Ken Olum, who is much more of a computer expert than Allen, was a bit more thorough. After some programming, Olum discovered that when he set $P = 0$ he didn't get the usual flat spacetime. Instead, he got a solution of the Einstein equations everywhere except at $r = 0$. There, on the axis, was a singularity, where the curvature of the spacetime became infinite. In fact, the singularity was always present *with or without* the light beam. It was just easier to recognize once the light beam was turned off. Moreover, this was not an artifact of the particular choice of coordinates. Furthermore, the singularity was *naked*, that is, it was not surrounded by an event horizon as in the case of a black hole. As discussed in appendix 6, naked singularities are a serious problem, because their behavior or what will come out of them cannot be predicted. They render the spacetime that they inhabit unpredictable.

The closed timelike curves disappeared when the light beam was turned off, so they were not entirely the result of the singularity; the light beam played at least a part. The strength of the singularity depended only on R_0. It could be eliminated only by letting R_0 become infinite. Since the closed timelike curves occur at $R > R_0$, one could not eliminate the singularity without also eliminating any possible time machine, even if one then turned the beam back on. This makes it impossible to say whether a light beam of finite radius without the singularity could cause the closed timelike curves, or whether they depend on having both the light beam and the singularity. The fact that, with just the singularity present, space is far from flat at infinity suggests that the closed timelike curves might be a cooperative result of the presence of both the light beam and the singularity, although this is not clear. However, the presence of the singularity in the absence of the light beam indicates that the spacetime is problematic to start with, even before the light source is turned on. In the Mallett model, we have a universe that, in principle, can have closed timelike curves, although unobservably far away. However, that universe is not the universe we live in because of the naked singularity.

Mallett's original article does not mention the problem of the singularity, nor is it addressed explicitly in his book, *Time Traveler*. There he says:

> I decided to dispense with trying to model mathematically either an optical fiber or a photonic crystal. Instead, for the sake of generality and to keep the light beam on a cylindrical path, I elected to use a geometric constraint. This constraint was represented by a static (nonmoving) line source. Light naturally wants to travel along a straight line. The only purpose of the line source in my calculations was to act as a general constraint to confine the circulating light beam to a cylinder. (Set up experimentally, the line source could look like wrapping a piece of string around a maypole, with the string being the light beam and the maypole serving as the line source.) The light beam itself would be conceived of as a massless fluid flowing in only one direction around the cylinder. This meant that the solution really contained two solutions: one for the circulating light and one for the static source.[3]

The idea, perhaps, is that the singularity approximates the gravitational field of the mirrors or light pipes that carry the laser beam.

It is important to note that, in fact, the problem that Mallett *actually solved* was one of an infinite cylinder of light with a line singularity on its axis, not a finite one where the light travels in light pipes. Furthermore, there is no reason to think that the effects of the line singularity along the axis that Olum uncovered, and those, say, of an infinitely long beam–carrying helical lucite light pipe wrapped around the axis would be good approximations to one another.

A recent article by Olum published in *Physical Review* shows that they are *not*, in fact, good approximations to one another at all. Using Mallett's equations, Olum calculated the paths of freely moving particles and light rays (timelike and lightlike geodesics, respectively) in Mallett's infinite cylinder model. He found that no timelike geodesic, or any geodesic (timelike or lightlike) that

3. R. L. Mallett with Bruce Henderson, *Time Traveler: A Scientist's Personal Mission to Make Time Travel a Reality* (New York: Basic Books, 2006), 167–68.

has any motion in the direction parallel to the cylinder axis, can escape to large distances. Furthermore, Olum found that every such geodesic originates and terminates in the singularity. Light rays moving entirely in the radial direction either outward or inward start or end, respectively, at the singularity.

For light rays moving perpendicular to the cylinder axis, there are several possibilities. There are some paths of light rays that orbit the singularity at fixed distance. These are the trajectories that Mallett found. More generally, light rays start at the singularity and gradually spiral out to arbitrarily large distances. Other light rays start at very large distances from the cylinder and spiral into the singularity, where they come to an end. Olum also calculated the behavior of a particle initially at rest some finite distance away from the singularity. He found that such particles fall into the singularity, and thus are destroyed, in a finite proper time. To quote Olum: "It therefore appears that any attempt to build a 'time machine' along the lines described by Mallett would have a very unfortunate effect on nearby objects."[4] Needless to say, this is not the behavior of particles and light rays in a system of light pipes in any sensible experimental setup.

In his book, Mallett claims that since the closed timelike curves are not present until the light source is turned on, it must be the light source that produces the closed timelike curves. However, given our earlier discussion, it is likely that without the singularity there would be no closed timelike curves. And even if there were, they would be at unobservably large distances.

Since this has been a rather long discussion, let us summarize. If one accepts the assumptions of Hawking's theorem (given in appendix 6), then it implies that no finite-sized time machine along the lines suggested by Mallett, using only classical matter (i.e., no negative energy), will ever be possible. Even ignoring this and taking Mallett's model at face value, one finds that it is fundamentally flawed for a number of reasons. The closed timelike curves predicted by the model only occur at distances that are unimaginable orders of magnitude larger than the visible universe. This is not merely a technological problem that will be remedied in the future by clever engineering using more powerful lasers. For a cylinder of 1-meter radius, even if one increased the laser power to the power output of all the stars in our galaxy, the closed timelike curves still occur at distances of $10^{\left(10^{11}\right)}$ meters, whereas the size of the visible universe is a mere 10^{26} meters. Another serious problem is that if one turns the

---

4. K. Olum, "Geodesics in the Static Mallett Spacetime." *Physical Review D* 81 (2010): 127501–3.

laser power off, Mallett's solution does not reduce to flat spacetime. Instead, on the axis of the cylinder there is a singularity, where the curvature of spacetime becomes infinite. Mallett claimed in his book that he used a "geometric constraint," that is, the line singularity, to model the apparatus that would hold the light in a circle in a more realistic setup. However, the behavior of the motion of particles and light rays in the vicinity of the singularity can be calculated using Mallett's equations. It is quite peculiar and certainly does not model the behavior of light rays in a system of light pipes.

We emphasize that what has been presented here is not an "alternative theory" to that of Mallett. Also, unlike the cases of other time machines we have discussed, our conclusions do not depend on appeals to the vagaries of some as yet unknown quantum theory of gravity. Mallett's model consists of classical general relativity with a classical matter source. This is an unambiguously solvable problem and, hence, a decidable question. The conclusions we have discussed are *direct consequences of the equations Mallett himself has presented* in his published paper.

Gott's Cosmic String Time Machine

The final example we'll discuss in this chapter is the "cosmic string time machine" discovered by Richard Gott of Princeton in 1991. Before turning to Gott's time machine, we will indulge ourselves in a brief discussion of cosmic strings, because they are fascinating objects in their own right. There also is a fair chance that they actually exist.

Cosmic strings are exceedingly thin but potentially incredibly massive objects. They have no ends, so either they occur in closed loops or they are of infinite length. Many elementary particle theories predict cosmic strings. Their possible existence in a cosmological context was suggested in a paper by Professor Tom Kibble of Imperial College, London. In most of these theories, the strings are so massive that, because of the Einstein relation $E = mc^2$, they can only be produced in the early universe—where very high-energy particles were present during a very short time following the big bang—and not at terrestrial accelerators.

If cosmic strings exist, then because of their large masses, they might very well be of considerable significance for cosmology and astrophysics. In addition, their predicted properties depend on the details of elementary particle theory. For that reason the discovery, or for that matter a definitive failure to discover, cosmic strings would give useful information about elementary particle theory in an energy range that is of great interest but is largely inaccessible experimentally. This question of what we may be able to deduce about what went on at the very high temperatures in the first tiniest fractions of a second of our universe thus provides a connection between physics at the very smallest scales, that is, particle physics, and at the very largest, that is, cosmology. Therefore, many theoretical physicists who, like Allen, considered themselves particle physicists, have found themselves becoming part-time cosmologists and general relativists.

A cosmic string is best described by giving its mass *per unit length*, which we will represent by m, the same symbol we used for the mass per unit length of the laser beam in our discussion of the Mallett proposal. Let T be the temperature at the time when a particular kind of cosmic string formed as the early universe cooled. We will give T in energy units, since it is the average particle *energy* in the universe at the time in question that is relevant here.

In the case of a Planck-scale string, T is about 10^{19} times the rest energy of a proton or about 10^{19} GeV. Here, the G stands for "giga," that is, billion. An "eV," or, "electron volt," is a basic unit of energy in particle physics. The string has a mass of about 10^{25} tons, or about 1,000 earth masses per meter. The width of such a string would be about the size of the Planck length, about 10^{-35} meters, or around 10^{29} times smaller than the radius of an atomic nucleus. Such a string would indeed be an amazing object, an incredibly small width combined with a mass of many tons in every meter of length.

However, there is a respectable theory in elementary particle physics, called grand unification, which would suggest that we might have cosmic strings that formed at a temperature of around 10^{15} GeV. Such a string thus would have an m of "only" about 10^{17} tons per meter.

There are a number of ways of searching for either direct or indirect observational evidence for cosmic strings, and none have been found, though there have been a couple of what seem to have been false alarms. Unfortunately, the sensitivity of all these methods is such that strings with the properties one might expect from particle theory are just on the verge of detectability. Thus, no firm conclusion can be drawn from the failure to detect them thus far, although things are getting a bit dicey for strings predicted by grand unified theories. Such strings are so massive that their existence can affect cosmological evolution. At one time there was great interest in the possibility that strings provided the "seeds" around which galaxies condensed out of the original homogeneous cosmic soup. However, it now seems that fluctuations associated with a very early rapid period of exponential expansion were responsible for galaxy formation. For an account of this subject you should see the splendid book *The Inflationary Universe*, by Alan Guth, who originated the idea of inflation.

Most of the methods for detecting strings depend on their gravitational properties. The nature of the spacetime around a long straight cosmic string is closely connected with Gott's scheme for a cosmic string time machine. The spacetime around a cosmic string was first elucidated by Allen's Tufts colleague, Alex Vilenkin, as a result of a chain of events in which Allen had a serendipitous involvement. Vilenkin solved the problem in an approximation in which the width of the string was neglected, which is in general a very reasonable approximation. Exact solutions were later given by several people, including Gott himself, and these confirmed Vilenkin's results.

Vilenkin grew up in Kharkov in the former Soviet Union. He went to Kharkov University, where he was very successful. However, because of his Jewish ethnicity, he found the path to graduate study was not open to him, and he wound up as a night watchman in the Kharkov zoo. Happily, the zoo was quiet during those hours, and Alex wrote and published two physics papers during that time. This was a time of détente in U.S.-Soviet relations, and some Jewish emigration was allowed. The Vilenkins were thus able to come to the United States, and Vilenkin entered the graduate program at the State University of New York at Buffalo, where he earned his doctorate in one year. After that, the Vilenkins moved two hundred miles or so along the Lake Erie shore, and Alex became, in 1977, a postdoc at Case-Western Reserve.

In 1978, Allen was grumbling his way through a term as chairman of the physics department at Tufts. Not being an administrator by taste, he had made a resoundingly Sherman-esque statement concerning his unwillingness to take the position when time came to choose a new incumbent. The fates, however, are sometimes fickle. A squabble developed between a majority of the physics department and the then dean of the faculty over who should be the next chairman. Allen found himself, seemingly, to be the person who was at least somewhat tolerable to everyone involved, and reluctantly decided to take the job.

To add to his problems, an unforeseen spring resignation led to the necessity of finding a fall replacement on rather short notice. As it happened, Allen heard that there was a young Russian named Alex Vilenkin who was currently a postdoc at Case-Western Reserve. Allen assumed that Vilenkin's letters of reference were exaggerated (it sometimes happens—however, these weren't) but it was something of an emergency, and so Vilenkin joined Tufts in the fall of 1978 with a one-year appointment as an assistant professor of physics. Allen and the other members of the physics department soon discovered that the United States—and Tufts in particular—owed the KGB a debt of gratitude unwittingly leading to the emigration of this young physicist of exceptional ability. Vilenkin's appointment was extended to a normal three-year term and then to an early grant of tenure. He has become one of the world's most eminent cosmologists and has written seminal papers in four different subareas of cosmology. He has been the director of the Tufts Institute of Cosmology since a private grant funded its establishment in 1989.

Cosmic strings are one of three classes of closely related objects called (generically, for reasons that are a bit too complicated to go into here) "topological defects." One of the other classes, and the first to be discussed, involves objects referred to as "domain walls," which, as the name suggests, are planar rather than string-like. Allen came across these during a sabbatical at MIT in 1973–1974 and wrote a rather minor paper that was one of the earliest ones on the subject to appear in *Physical Review*. In 1979 he wandered into Vilenkin's office with a question he had been thinking about concerning the gravitational effects of a domain wall; he brought with him a copy of Kibble's paper, mentioned earlier, discussing both cosmic strings and domain walls.

Allen's question turned out to have a rather complicated answer, and it was about a year before Vilenkin produced a paper with an answer, as well as, for good measure, his discussion of the gravitational effect of a long straight cosmic string. Vilenkin went on to become perhaps the world's leading expert on topological defects. In addition to his many papers on the subject, on some of which Allen has collaborated, he has written, together with Paul Shellard of the University of Cambridge, the definitive book in the field. One of Allen's prized possessions is a copy of *Cosmic Strings and Other Topological Defects* with Vilenkin's signature, including a note of thanks for asking that question that got Alex interested in the subject.

What Vilenkin discovered about the spacetime outside a long straight cosmic string is rather remarkable. The space is flat, that is, without curvature, and hence, despite its huge mass, a straight cosmic string exerts no gravitational force on surrounding objects. Thus the space around such a string at a given moment of time could be drawn on an undistorted flat sheet of paper on which the usual Euclidean geometry holds, the sum of the angles of a triangle, for example, would be 180° or π radians. However, a path around a circle of radius r centered on the string, at constant time in the reference frame in which the string is at rest, will have a length of $(2\pi - \theta)r$, rather than $2\pi r$ as expected from the familiar formula for the radius of a circle. It is as though someone

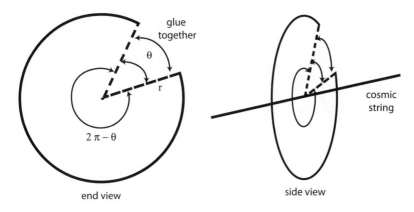

FIG. 13.2. Space in the vicinity of a cosmic string.

had taken a scissors and cut a wedge-shaped piece of pie of angle θ out of the paper and then glued the two sides of the wedge together, as illustrated in figure 13.2, so that simultaneous events on opposite sides of the wedge become identified with one another.

In the cosmic string case the resulting space is called a "conical" space, since in order to glue the two sides of the wedge together, the sheet of paper must be deformed into a cone. The paper remains locally flat everywhere in this process, just as it would if we rolled the paper up into a cylinder. The angle θ is called the "deficit" angle. It is determined by the mass per unit length, m, of the string and is given by $\theta = 8\pi(G / c^2)m$, where G is Newton's constant, as long as θ is not close to 2π.

It should be emphasized that once the sides of the wedge are glued together, you can no longer tell where the wedge was. You would feel no jolt as you cross it, if you were to fly around the string in a spaceship. In describing the situation you can take the wedge to be anywhere you find convenient just by an appropriate choice of where you take the angle $\theta = 0$ when you choose your coordinate system.

By making use of cosmic strings and their properties, Gott found a very clever way of producing a closed timelike curve. To see how this comes about we will make use of the following argument, which is slightly different than Gott's. We first note that the missing wedge has the effect of allowing super-luminal travel. Now consider a case where you have an infinite straight string along the z axis whose position in the xy plane, in the inertial frame in which the string is at rest, is just off the x axis at $y = e$ where $y = e$ is taken to be much

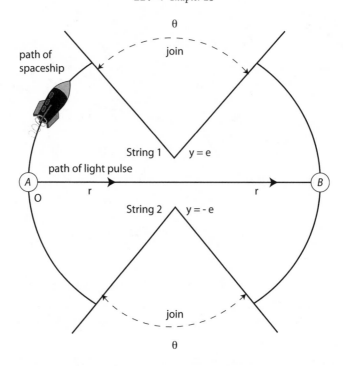

FIG. 13.3. Gott's cosmic string time machine. The two strings are perpendicular to the page. String 1 moves to the left, while String 2 moves to the right. By starting at A and moving first around String 1, and then around String 2, along the path shown, a space traveler can return to the same point in space and time.

less than r (refer to the top half of figure 13.3). We take the tip of the wedge, with deficit angle θ, to be at $x = r, y = e$, and oriented as shown in figure 13.3. Suppose that at time $t = 0$ you send a light pulse from planet A, located at the origin along the x axis, to planet B located at $x = 2r$ so that the signal arrives at $t = 2r/c$. The signal will pass close by the string, but be unaffected by it since the wedge is on the opposite side of the string. (Again we emphasize that the results do not depend on where we take the wedge. Our choice just makes the algebra and geometry easier to visualize.) At the same time a spaceship, traveling at nearly the speed of light, sets off for planet B, following a semicircular path enclosing the string.

The center of the path is halfway between the planets and thus almost centered at the string, since e is very small. If it weren't for the missing wedge, the spaceship would travel a distance πr along its path and arrive after the light pulse, which was slightly faster, but, more importantly, followed the shorter, straight-line path. However, the spaceship just jumps from one edge of the wedge to the other, and travels only a distance $(\pi - \theta)r$. If θ is appreciably greater than $\pi - 2$, the linear distance covered by the spaceship will be less than that covered by the light pulse,[5] and the spaceship will beat the light pulse to planet B, arriving at time $t_{ship} = \dfrac{(\pi - \theta)r}{c} < t$. (We remind the reader that we are assuming that the spaceship travels at very nearly the speed of light. Hence, we have approximated its velocity here by c.)

Since the ship has covered the same distance between A and B as the light pulse in less time, it has, in effect, undergone superluminal travel, traveling along the x axis at an effective speed $u = 2r / t_{ship} > c = 2r / t$. This is a situation we have encountered before. We know that spacetime intervals along the world-line of a superluminal object are spacelike, and the sign of the time component is not Lorentz invariant. The spaceship travels forward in its own proper time, along a spacelike path. However, because the time *order* of two events which are spacelike separated is not invariant, one can find a Lorentz frame where the two events occur simultaneously. Gott showed that if the string moves along the x axis relative to the frame in which the planets are at rest, then in that frame the arrival of the spaceship at planet B can occur simultaneously with its departure from planet A. The scenario is reminiscent of our discussion of tachyons back in chapter 6.

However, we've not yet built a time machine. To do that we must arrange for the spaceship to return to its starting point in space and time so that it can affect its own past. That can't be done with only a single cosmic string. If the spaceship just retraces its path along the semicircular curve, it turns out that no time machine is possible.

Professor Gott had a clever idea, however. He considered a pair of infinite parallel cosmic strings, each moving at a speed very close to the speed of light and perpendicularly to its length, but in *opposite* directions, so that they zoomed past one another. Let's call them string 1, moving on the path with $y = e$, and string 2, at $y = -e$. The vertices for the two wedges were at the respective strings, and the wedges were oriented in the positive and negative

---

5. This is just the condition that the distance along the curved path taken by the spaceship be less than the straight line distance taken by the light pulse, i.e., $(\pi - \theta)r < 2r$. A slight rearrangement then gives $\pi - 2 < \theta$.

y directions, respectively. (Refer to the entirety of figure 13.3; string 1 moves to the left and string 2 moves to the right.)

Gott showed that one can now get a time machine by sending the spaceship around a closed path where it passes through the missing wedge for string 1 on the way from A to B, and the wedge for string 2, moving in the opposite direction, on the return trip. It can be arranged so that the spaceship arrives back at planet A simultaneously with its departure. The spaceship has now returned to the same point in space and time, that is, it has traveled on a closed timelike curve. This leaves open, at least in principle, the possibility of a time machine in which the spaceship travels into its own past. The scenario of the two moving cosmic strings is analogous to the case of two tachyon transmitters moving relative to one another.

Gott's paper discussing this was published in Physical Review Letters. The editors of this very prestigious journal attempt to restrict publication to articles that they feel are so important they deserve especially rapid publication; for that reason, most of us in the profession are quite pleased if one of our papers gets accepted there. And the editors probably receive a number of forceful—or even vituperative—dissents from authors who have one of their pet papers turned down.

There were no such problems with Gott's paper, which almost everyone would agree deserved its place in Physical Review Letters. Allen not only had a chance to read Gott's paper but to hear him deliver an early lecture on the subject when he accepted an invitation to speak at the weekly Boston area cosmology seminar that rotates between Tufts, Harvard, and MIT. The Tufts cosmology group particularly looked forward to his lecture because of the extensive research on cosmic strings and their potential importance in cosmology that was being conducted at Tufts, and because Allen, particularly, had a history of interest in the physics of time travel. Gott did not disappoint, delivering an excellent lecture in terms of both content and presentation.

Since the energy density of cosmic strings is positive, they are not "exotic" in the technical sense, though you might well think they are somewhat bizarre. As in the previous models, the cosmic strings in Gott's time machine escape the strictures of Hawking's theorem only because of their infinite length. (In particular, Hawking's theorem forbids the construction of a time machine using finite loops of cosmic string.) However, Gott's idea prompted more interest, since theory suggested that, although one could not manufacture an infinitely long cosmic string on demand, still, the "ingredients" for a time machine might exist. In this situation, even though they are being produced at a finite time after the big bang, the fact that they cannot end will force some of them to have infinite length (provided that the universe is infinite in size, as suggested by current measurements). If strings were produced in the very early universe, a few at least could still be around. According to theory, there might now be only a small number in our whole visible universe. The chances of two of them being, randomly, in the right relation to one another to produce a time machine doesn't seem promising, to put it mildly. Moreover, they must be produced with very high speeds, and therefore require a lot of kinetic energy in addition to their intrinsic mass-energy per unit length. Even if they are possible in principle, looking for a Gott time machine might be like waiting around at the swimming pool to see a diver pop spontaneously out of the water.

14
Epilogue

"The time has come," the walrus said,
"To talk of many things."

> LEWIS CARROLL, *Through the Looking-Glass
> and What Alice Found There*

If you can look into the seeds of time,
And say which grain will grow, and which will not,
Speak.

> WILLIAM SHAKESPEARE, *Macbeth*

Since we have now come to the end of our journey through time and space, let's summarize where we've been, where we are, and what the prospects are for the future. We've seen that Einstein's equations of general relativity seem to allow for the possibility of faster-than-light shortcuts and backward time travel. However, we've also seen that there appear to be severe restrictions on the actual realization of wormholes, warp drives, and time machines, especially when we consider the laws of quantum mechanics. Given the existing research, our view is that the construction of such objects seems to be extremely unlikely, at least in the forms suggested to date. This is a rather depressing conclusion if we someday wish to cross the enormous gulf of space between the stars, and "boldly go where no one has gone before." But how trustworthy can our conclusions be, given our present state of knowledge? How well can we predict twenty-third-century possibilities, on the basis of a twenty-first-century knowledge of physics? Might we not expect future discoveries to overturn even some theories that we presently regard as firmly established, as has happened often in the history of science? Here we offer a few relevant speculations.

< 218 >

An efficient method of space travel could be an important issue for the survival of the human race. For example, we know that asteroid impacts have occurred numerous times in the history of our planet. One such impact sixty-five million years ago quite probably ended the reign of the dinosaurs. We know that if we remain on this planet long enough, eventually another such catastrophic impact will happen and possibly herald the end of our species. But it could happen a million years from now, or in the next ten years; we just don't know. If we stay on only one planet, we risk annihilation due to this or some other global catastrophe. So it would seem that it should be a fundamental goal for us to develop the capability to get off the planet (and out of the solar system).

On the other hand, consider the times in the earth's history when technologically advanced societies have come into contact with less advanced ones. The outcome has usually not been a happy one for the latter. Hence, from one point of view, the huge distances between the stars and the technological obstacles to quick interstellar travel could be a blessing rather than a curse. It might prevent the aggressive inhabitants of the galaxy (including us—the *Star Trek* "Prime Directive" notwithstanding) from wreaking havoc on the peaceful ones.

What is the likelihood that the conclusions we have reached will stand the test of time, particularly the "no-go" results? This is difficult to say, but we can make some informed guesses. We have seen over and over again in the history of physics how new theories have replaced earlier ones. Relativity and quantum mechanics replaced Newtonian mechanics with an entirely new worldview. Might not similar revolutions overthrow our current conclusions? However, it is important to remember that *the new theories must agree with the old theories in the regime where the old theories are known to agree with experiment.* Relativity and quantum theory reduce to Newtonian mechanics for weak gravitational fields, speeds that are small compared to that of light, and sizes that are very large compared to those of microscopic objects. We expect deviations from the old theories only in domains where the original theories are no longer applicable.

When applying the quantum inequalities to wormhole and warp drive spacetimes, we assume that the flat spacetime inequalities can be used in curved spacetime, if we restrict our sampling time to be small compared to the radius of curvature of the spacetime and the distance to any boundaries. On that scale spacetime is approximately flat and one "doesn't notice" the curvature or the presence of boundaries. This is a reasonable assumption in accord with the principle of equivalence in general relativity. That is like saying that in order to reliably predict the outcome of a local laboratory experiment on earth, we don't need to know that spacetime is curved on the large-scale or that there might be a boundary (e.g., like a Casimir plate) many light-years away. Were that not the case, we would notice deviations from the incredibly accurate predictions of quantum field theory on laboratory scales, which in fact we don't observe. It is hard to see how the validity of our "locally flat" assumption would be called into question at some time in the future. Given that, then we are essentially just using experimentally tested quantum field theory on the scales where we know it to be true, in order to obtain our bounds on wormholes and warp drives. So it's rather hard to see how to avoid (possibly large) negative energy that is restricted to very small regions of space or time.

So if there are ways around our conclusions, what might they be likely to entail? The quantum inequalities are strong restrictions, but they have been proven to hold for free fields. If no such strong restrictions exist for interacting quantum fields, then that might be a way around our conclusions. Although the situation is still rather murky as of this writing, we feel it unlikely that this will be the case. Our bet is that while interacting fields might not obey the usual quantum inequalities, they probably satisfy some kind of similar constraints. However, at present we have no proof.

Another possibility is that the dark energy that drives the accelerated expansion of the universe turns out to be exotic material, in the sense of violating the weak or null energy conditions (or the corresponding averaged conditions). Then we could have exotic matter all around us. Although, again as of this writing, this is not ruled out by observation, we would be quite surprised if it were the case. On the other hand, the very existence of dark energy came as a big surprise to most physicists and astronomers.

As for time travel, it seems that, in view of the "slicing and dicing" effect of wormhole time machines in the many worlds interpretation, one is stuck with something like the banana peel mechanism if one is to avoid time travel paradoxes. However, we find it extremely disturbing that, in order to preserve the laws of quantum mechanics, the construction of a time machine in the future can affect one's ability to accurately predict the results of experiments in the present! Not much work is done on time travel these days, because the general feeling in the relativity community is that we've gone about as far as we can go without a quantum theory of gravity.

One does have to be careful in making predictions about future theories and technologies on the basis of present ones. Consider the following example (provided by Ken Olum). Suppose we were given only the laws of Newtonian mechanics and no new technology. We would then probably say that it is physically impossible, even in principle, to travel interstellar distances in a human lifetime. Humans could not survive the accelerations needed to cover such distances in their lifetimes, given the assumptions of absolute space and time in Newton's theory. However, that conclusion would be wrong.

With the advent of special relativity, and the phenomenon of time dilation, we learned that the passage of time on a spaceship traveling near the speed of light can be vastly different from the passage of time on earth. As for the accelerations required to reach these speeds, in principle it would be possible to accelerate at a constant acceleration of 1g for a year or two, in order to achieve near-light speeds. We emphasize here that the *laws of physics* do not prevent you from traveling as *close* to the speed of light as you like. Relativistic time dilation does, in principle, provide a way of bridging the distances between the stars.

Of course, *in practice*, there are a whole lot of other problems, for example, you are generally stuck coming back a long time after everyone you know is dead. That makes it a bit tough to organize any kind of galactic federation, unless you are a very patient and long-lived species. One possibility might be to send robots instead. Another problem is that once your starship has been accelerated up to near-light speed, you have a big shielding problem to worry about. In the frame of the ship, you are at rest while interstellar atoms and dust whizz past (and through!) you at enormous speeds. For you, it's like sitting in the middle of a particle accelerator. The protective shielding needed would likely dramatically increase the mass of your ship.

The previous discussion has centered on the difference between what the laws of physics allow and what is possible in engineering. But people yearn for a *quick* way to get to the stars. In this book, we've argued that travel to the stars via some sort of *superluminal* travel, the way people are used to seeing it in science fiction, is what appears to be problematic.

Another point (credited to Doug Urban of Tufts) is that it is possible for us to make quantum-mechanical matter and energy in the laboratory, whose properties we might not have guessed based solely on the laws of classical physics. Examples are liquid helium, which can crawl up the walls of its container; Bose-Einstein condensates, a new state of matter that can exhibit strange quantum mechanical behavior on a macroscopic scale; and lasers, which are now routinely used in many areas of our technology. Although the laws of quantum mechanics and relativity reduce to those of Newtonian physics at large scales and low velocities, the former can still be used to produce effects that are tangible on human scales.

What kind of frontiers might lie ahead? Physicists are currently investigating higher energies and the related probing of ever-smaller scales of space and time. Currently, two leading candidates for a quantum theory of gravity are string theory and loop quantum gravity.[1] A theory of quantum gravity could, and many believe would, be as scientifically revolutionary as quantum mechanics, but will it affect humanity to the same extent? The energy scale of quantum gravity is so enormous that we may not be able to manipulate its effects in the near future, if ever.

However, if that is not the case, we might *imagine* the following scenario. The unknown laws of quantum gravity will presumably describe, among other things, the behavior of large amounts of matter compressed into almost inconceivably tiny regions of space. Perhaps these laws incorporate natural ways to effectively circumvent or supersede energy conditions. (We might expect the laws of quantum gravity to have this property if we believe that they will ultimately resolve the problem of singularities in spacetime.)

Imagine, for example, a super-civilization manipulating quantum-gravitational matter and energy into long string-like negative energy configurations, which might even satisfy the demands of the quantum inequalities. These negative energy-type strings might serve as the source of exotic matter for building one of Matt Visser's cubical wormholes (discussed in chapter 9). Recall that one of the advantages of this type of cubical wormhole is that exotic matter is confined to the edges of the cube. This means a human observer could pass through a face of the cube and through the wormhole without ever directly encountering the exotic matter. Such a device could provide a gateway to the stars. Would the laws of quantum gravity then allow us to combine many of these together to make a "Visser ring" of wormholes into a time machine? If the laws of quantum gravity do not allow wormhole time machines, might they allow other types? Or do these laws forbid time machines altogether, enforcing Hawking's chronology protection conjecture? At this juncture we don't

1. For more on these theories, see Brian Greene, *The Elegant Universe* (New York: W. W. Norton, 2003); Lee Smolin, *Three Roads to Quantum Gravity* (New York: Basic Books, 2001); and Lee Smolin, *The Trouble with Physics* (Boston: Houghton Mifflin, 2006). A nice article on loop quantum gravity is Lee Smolin, "Atoms of Space and Time," *Scientific American*, January 2004.

know. But let us emphasize that the discussion in the last two paragraphs is just *pure speculation*. Presently, we have no reason to believe the above scenario is possible.

By this time, you may feel that your authors are just old curmudgeons who just want to ruin everyone's fun.[2] However, it might surprise you to learn that we are both avid *Star Trek* fans. Like you, perhaps, we think that the universe might be much more exciting with the existence of wormholes and time machines. But it's precisely because we feel that way that we are cautious. We adopt the maxim that one should be most skeptical about that which one would most like to believe. As Richard Feynman once put it, "The first principle is that you must not fool yourself—and you are the easiest person to fool." We also abide by Carl Sagan's famous quote, "Extraordinary claims require extraordinary proof"—and the burden of proof lies with the claimant. As scientists, it's our job to understand the universe as it is, not as how we might wish it to be. We must always keep in mind that the universe is under absolutely no obligation to fulfill our hopes and desires. However, we would argue that, in any case, the new insights that have been gained about time and space, matter and energy, have made our journey worthwhile.

2. In fact, we have already been so accused by E. W. Davis and H. E. Puthoff (in CP813, *Space Technology and Applications International Forum—STAIF 2006*, edited by M. S. El-Genk [Melville, NY: American Institute of Physics Press, 2006]): "The Quantum Inequalities (QI) Conjecture is an ad hoc extension of the Heisenberg Uncertainty Principle. [Authors' note: Not true—the quantum inequalities are derivable from quantum field theory, so they are not a "conjecture."] They were essentially derived by a small group of curved spacetime quantum field theory specialists *for the purpose of making the universe look rational and uninteresting* [emphasis added] . . . This small group is prejudiced against faster-than-light motion, traversable wormhole and warp drive spacetimes, time machines, negative energy, and other related issues having to do with the violation of the second law of thermodynamics. This group accepts the reality of the theoretical and proven experimental existence of negative energy density and fluxes, but they don't accept the consequences of its various manifestations in spacetime." It should be mentioned that this is the same H. E. Puthoff who, together with Russell Targ, declared Uri Geller a genuine psychic in the 1970s. If you have never heard of Uri Geller, we suggest *The Truth about Uri Geller*, by James ("the Amazing") Randi (Amherst, NY: Prometheus Books, 1982.) We apologize for exploring the hypothesis that the universe looks rational. It appears that this is a concern that Davis and Puthoff do not share.

Appendix I
Derivation of the Galilean Velocity Transformations

The Galilean velocity transformations can be easily gotten from the Galilean coordinate transformations in the following way. Recall from chapter 2 that the latter are given by

$$x' = x - vt$$
$$y' = y$$
$$z' = z$$
$$t' = t.$$

Now suppose a person walks from a point with coordinate x_1' in the train's reference frame to the point with coordinate x_2' in the time interval between t_1' and t_2'. Thus, x' changes by $x_2' - x_1'$ while t' changes by $t_2' - t_1'$. We will make use of a commonly used notation with which many readers will be familiar. We represent $x_2' - x_1'$ by the symbol $\Delta x'$. The symbol Δ is the Greek letter Delta, and $\Delta x'$ is read as "Delta x prime," or, the "the change in x prime." That is, $\Delta x'$ is a single symbol that is just a convenient shorthand for the quantity $x_2' - x_1'$; it is *not* x' multiplied by some mysterious quantity Δ. Similarly $\Delta t'$ represents the change in t'. Thus, $\Delta t' = t_2' - t_1'$.

Let Δx be the distance the person moves in the time interval Δt (recall that we assume to be equal to $\Delta t'$) along the track. For generality, let us include the possibility that the person also moved in the other two directions, say by amounts $\Delta y' = \Delta y$ and $\Delta z' = \Delta z$. Then the Galilean coordinate transformations give us

$$\Delta x' = \Delta x - v\Delta t$$
$$\Delta y' = \Delta y$$
$$\Delta z' = \Delta z.$$

Now simply divide the left-hand side of each equation by $\Delta t'$, and divide the right-hand sides by Δt (remember that $\Delta t' = \Delta t$, so we are really dividing both sides by the same quantity). Then we get

< 225 >

$$\frac{\Delta x'}{\Delta t'} = \frac{\Delta x}{\Delta t} - \frac{v\Delta t}{\Delta t}$$

$$\frac{\Delta y'}{\Delta t'} = \frac{\Delta y}{\Delta t}$$

$$\frac{\Delta z'}{\Delta t'} = \frac{\Delta z}{\Delta t}.$$

But $\frac{\Delta x'}{\Delta t'}$ is just the distance the person walks on the train divided by the time interval as measured on the train, that is, the speed u' relative to the train. Similarly $\frac{\Delta x}{\Delta t}$ is the distance the person travels with respect to the track in the time interval Δt, the time interval as measured by clocks in the track frame (assumed to be the same in the train frame), that is, the speed u relative to the track. So we have that

$$\frac{\Delta x'}{\Delta t'} = \frac{\Delta x}{\Delta t} - \frac{v\Delta t}{\Delta t}$$

$$u' = u - v.$$

Recall that u', u represents only the parts of the motion that are parallel to the track. If the person moves in the other two directions as well, then using a similar argument to the one above, we have that

$$\frac{\Delta y'}{\Delta t'} = \frac{\Delta y}{\Delta t}$$

$$V'_y = V_y$$

$$\frac{\Delta z'}{\Delta t'} = \frac{\Delta z}{\Delta t}$$

$$V'_z = V_z,$$

Where V_y', V_y, V_z', V_z are the velocities in the y', y, z', z directions, respectively.

Appendix 2
Derivation of the Lorentz Transformations

We are going to derive the Lorentz transformation equations, using a derivation that is essentially the same as one originally given by Einstein. As you know, the transformations give the coordinates of an event in a reference frame S' in terms of the coordinates in a different inertial frame S. As usual we will suppose that S' is moving in the positive direction along the common x and x' axes with speed v, and the origins of the two frames coincide at $t = t' = 0$. For the moment we will take the event to occur on the x axis so that its location in spacetime is specified by coordinates (t,x) in S and (t',x') in S'. To satisfy the principles of relativity, the transformation equations must ensure that a light signal moving in the positive direction with worldline given by $x = ct$, or $x - ct = 0$ in S moves along the worldline $x' - ct' = 0$ in S'. This will be true if the transformation equations are such that

$$x' - ct' = a(x - ct). \tag{1}$$

Here, a is a constant. That is, it does not depend on any of the coordinates in the equation, although it will depend on v. Equation 1 guarantees that $x' - ct' = 0$ if $x - ct = 0$, as long as a is not infinite. The principle of relativity also requires that a light signal moving in the negative direction in S, that is, along the worldline $-x = ct$ or $x + ct = 0$, have speed c as seen in S'. This can be guaranteed if we require that

$$x' + ct' = \beta(x + ct), \tag{2}$$

where again β is independent of the coordinates, and β is not infinite. We can introduce two more convenient constants, a and b. We first add equations 1 and 2 to get

$$x' = ax - bct \tag{3}$$

where

$$a = \frac{a + \beta}{2} \tag{4}$$

< 227 >

and

$$b = \frac{a - \beta}{2}. \tag{5}$$

We then subtract (2) from (1) to obtain

$$ct' = act - bx. \tag{6}$$

If we look at equations 3 and 6, we see that finding a and b will solve our problem. They are the two coefficients in the transformation equations that will allow us to determine the coordinates of an event in S' in terms of its coordinates in S, that is, the coefficients in the Lorentz transformations.

To make some further progress, let's observe that the origin of S' is located at $x' = 0$. From equation 3, its position in S is thus at $x = \frac{bc}{a}t$. Since it starts at $x = 0$ and moves with speed v relative to S, its position in S is also given by $x = vt$. Comparing the two expressions for x we see that

$$v = \frac{bc}{a}. \tag{7}$$

Next, consider a meter stick at rest in S', with one end at $x' = 0$ and one end at $x' = 1$ m. Let's find its length as measured by observers in S. Since the meter stick is moving, to determine its length they will have to be careful to measure the position of its two ends at the same time, which, of course, for them means the same value of t. One end of the meter stick is at the origin of S', which we know passes the origin of S at $t = t' = 0$. So to find the length of the meter stick in S we have to find out where the point $x' = 1$ m is when $t = 0$. That's easy to do. We can just look at equation 3 and see that, when $t = 0$ and $x' = 1$ m,

$$x = (1\text{m}) / a. \tag{8}$$

The fact that $x \neq 1$ is an example of one of the well-known consequences of special relativity, namely, that moving meter sticks appear shortened; this is discussed in more detail in appendix 5. But this is relativity, where all inertial frames, and in particular S and S', are created equal. Therefore, if the transformation equations are to respect the principles of relativity, they must ensure the same thing happens for a meter stick running from $x = 0$ to $x = 1$ m when it is observed in S'. Observers in S' will say that, in order to make a correct measurement of the length of what is, for them, a moving meter stick, they must measure the position of its two ends simultaneously, that is at the same value of t'.

This time we have to do a little more work. We know the two origins pass one another at $t = t' = 0$. So at $t' = 0$, one end of the meter stick in S will be at $x' = 0$. Equation 6 tells us that at $t' = 0$,

$$x = act/b. \tag{9}$$

But we don't know t. However, we can eliminate t by using equation 3 to write $t = (ax-x') / bc$. If we substitute this expression for t into equation 6, with $t' = 0$, we obtain $x = (a^2/b^2)(x-x'/a)$. In this expression, collect the terms with x on one side, multiply both sides by $-b^2/a^2$, and note from equation 7 that $b^2/a^2 = v^2/c^2$. One is then left with $x'/a = [1-(v^2/c^2)]x$ or $x' = a[1-(v^2/c^2)]x$. Then, multiplying the right side by a/a, we get

$$x' = a^2[1-v^2/c^2](1m)/a, \tag{10}$$

since $x = 1$ m at the other end of the meter stick with one end at the origin of the unprimed system. Now compare equations 8 and 10. Because a meter stick in the unprimed system should look the same to observers in the primed system as the other way around, the principles of relativity require that equation 1 must be identical to equation 8, with x replaced by x'. We conclude that $a^2(1-v^2/c^2) = 1$, or

$$a = \frac{1}{\sqrt{1-v^2/c^2}}. \tag{11}$$

Thus, we have determined one of the two coefficients appearing in the Lorentz transformation equations for x and t. The other coefficient, b, is then immediately given, in terms of a, by equation 7, as

$$b = va/c = \frac{1}{\sqrt{1-v^2/c^2}} \frac{v}{c}. \tag{12}$$

If you substitute equations 11 and 12 into equations 3 and 6, you will recover the Lorentz transformation equations for x and t given in chapter 3.

Adding and subtracting equations 4 and 5, we have that $\alpha = a+b$ and $\beta = a-b$. Since $v < c$, neither a nor b is ever infinite, and hence, neither is α nor β. Thus, it follows from equations 1 and 2 that the transformation equations indeed guarantee that if a particle is moving in the positive or negative x direction in S with speed c, this is also true in S'. That is, the Lorentz transformations equations do indeed leave the speed of light invariant. However, since neither α nor $\beta = 1$, then unless $ct-x = 0$, we will have $ct'-x' \neq ct-x$, and similarly for

$ct + x$. However, $\alpha\beta = (a+b)(a-b) = a^2 - b^2 = 1$, as can be easily confirmed from equations 11 and 12. We can write $((ct')^2 - x'^2) = (ct' - x')(ct' + x') = \alpha\beta \, (ct - x)$ $(ct + x)$, from equations 1 and 2. Since $\alpha\beta = 1$, therefore, the equation

$$(ct')^2 - x'^2 = (ct)^2 - x^2 \tag{13}$$

is valid for any values of t and x.

So far, we have considered only light signals propagating along the x axis. To discuss signals propagating in arbitrary directions, we must introduce the transverse coordinates, y and z. Since S' is moving in the x direction, there is no reason these coordinates should be any different in S' than in S, and so we take the last two members of the set of Lorentz transformations to be

$$y' = y \tag{14a}$$

and

$$z' = z \tag{14b}$$

Let's now consider a light pulse emitted from the origin at $t = 0$ in S in an arbitrary direction. Its position at time t will be given by

$$x^2 + y^2 + z^2 - (ct)^2 = 0, \tag{15}$$

where

$$r = \sqrt{x^2 + y^2 + z^2} \tag{16}$$

is the spatial distance from the origin in the inertial frame S. The Lorentz transformations imply that $x^2 - (ct)^2 = x'^2 - (ct')^2$. Together with equations 14a and 14b, this allows us to rewrite equations 15 as

$$x'^2 + y'^2 + z'^2 - (ct')^2 = 0.$$

We see that the Lorentz transformations guarantee that the light pulse also propagates outward in the radial direction with speed c in S' as required by the principles of relativity.

Recall that we took α and β in equations 1 and 2 to be constants, that is, independent of the spacetime coordinates. It follows that the coefficients in the transformation equations, which are constructed from α and β, are likewise constants, meaning that the coordinate transformations are *linear* equations. It thus follows from Einstein's derivation that they are the *unique* set of linear equations consistent with the principles of relativity.

However, what if one allows α and β to be coordinate dependent? Equations 1 and 2 would still guarantee that the velocity of light was the same in all inertial frames, because of the factors of $x \pm ct$ on the right sides of the equations. Is there any reason to prefer linear transformation equations?

We now have one very good reason for doing so—in fact, it is the best of all reasons for believing any physical theory. There is an immense body of experimental data from high- energy physics experiments supporting the validity of the principles of relativity with the linear Lorentz transformation equations incorporated. But of course these experiments had not yet been done at the time when Einstein was developing special relativity.

However, there was a compelling argument that, one suspects, caused Einstein to take the linearity of the transformation equations more or less for granted. There is a fundamental assumption that the laws of physics are the same everywhere and at all times. Physicists phrase this by saying that physical laws are symmetric under translations, that is, under displacements of the coordinate system, in either time or space.

What are the grounds for assuming the existence of these symmetries? It is certainly the simplest assumption to make, and perhaps the most aesthetically pleasing. But while it often seems to be true, we have no guarantee that nature will choose either to be simple or to appeal to human aesthetics.

In particular, the laws of conservation of momentum and of energy, probably the two most familiar conservation laws, can be derived just from the assumptions of invariance under translations in space and time, respectively. The existence of those two great conservation laws thus provides powerful evidence that physical laws do not single out any particular region of space or time as being different from any other. If that is true, the coordinate transformation equations should be linear, and thus, the Lorentz transformations are singled out from any other possibilities.

Appendix 3
Proof of the Invariance of the Spacetime Interval

Note: You already know this if you worked through appendix 2, where we derived the Lorentz transformations. Here we give the simpler argument needed if one assumes the transformation equations to begin with.

We verify that the following equation is true for the coordinates of a given event in two different inertial frames with relative velocity v:

$$x^2 - (ct)^2 = x'^2 - (ct')^2.$$

We use the Lorentz transformations:

$$t' = \frac{t - \dfrac{vx}{c^2}}{\sqrt{1 - \dfrac{v^2}{c^2}}}, \quad x' = \frac{x - vt}{\sqrt{1 - \dfrac{v^2}{c^2}}}, \quad y' = y, \, z' = z.$$

To begin, substitute the expressions for x' and t' in terms of x and t from the Lorentz transformations into the first equation.

$$x^2 - \left(ct\right)^2 = \left(\frac{x - vt}{\sqrt{1 - \dfrac{v^2}{c^2}}}\right)^2 - \left(c\left(\frac{t - \dfrac{vx}{c^2}}{\sqrt{1 - \dfrac{v^2}{c^2}}}\right)\right)^2$$

You must then carry out the process of squaring the resulting expressions, making use of the good old result for the square of a binomial, $(a + b)^2 = a^2 + 2ab + b^2$, and gather like terms together over the common denominator, $\dfrac{1}{1 - \dfrac{v^2}{c^2}}$.

$$x^2 - \left(ct\right)^2 = \left(\frac{x^2 - 2xvt + v^2t^2}{1 - \dfrac{v^2}{c^2}}\right) - c^2\left(\frac{t^2 - 2t\dfrac{vx}{c^2} + \dfrac{v^2x^2}{c^4}}{1 - \dfrac{v^2}{c^2}}\right)$$

< 232 >

$$x^2 - \left(ct\right)^2 = \frac{x^2 - 2xvt + v^2t^2 - c^2t^2 + 2tvx - \left(\dfrac{v^2}{c^2}\right)x^2}{1 - \dfrac{v^2}{c^2}}$$

Next cancel the $-2xvt$ and $2xvt$ terms, and collect the x^2 and the t^2 terms, in the numerator.

$$x^2 - \left(ct\right)^2 = \frac{x^2 - \left(\dfrac{v^2}{c^2}\right)x^2 + v^2t^2 - c^2t^2}{1 - \dfrac{v^2}{c^2}}$$

Now factor the x^2 and the t^2 terms, pulling out a factor of $-c^2$ for the latter terms:

$$x^2 - \left(ct\right)^2 = \frac{\left(1 - \dfrac{v^2}{c^2}\right)x^2 - \left(1 - \dfrac{v^2}{c^2}\right)c^2t^2}{1 - \dfrac{v^2}{c^2}}.$$

Finally, if we cancel out the factors of $\dfrac{1}{1 - \dfrac{v^2}{c^2}}$ out of the numerator and denominator, we have

$$x^2 - (ct)^2 = x^2 - c^2\,t^2.$$

So you find that the right-hand side of the original equation has reduced to the corresponding equation in the earth frame, that is, the left-hand-side. Since the equation $x^2 - (ct)^2 = x'^2 - (ct')^2$ is true in general, it is true when the quantities on the left and right side both equal zero, which you may recall is the equation for a light ray. So, if the coordinates in 2 different reference frames are related by the Lorentz transformations, then a signal that moves with speed c in one frame is also seen to move with speed c by observers in the other. Thus, observers in all inertial frames will observe that light travels at speed c relative to their own inertial frame, that is, the inertial frame in which they are at rest. This agrees with the outcome of the Michelson-Morley experiment and implies that the seemingly obvious Galilean transformations are actually an approximation—albeit a very good one—for speeds much less than c.

Appendix 4

Argument to Show the Orientation of the x', t' Axes Relative to the x, t Axes

In figure A4.1, let ct', x' be the spacetime axes of an observer O' moving with speed v relative to the observer O in the frame with the axes ct, x. We are going to show that the primed axes are both rotated inward (i.e., toward the worldline of the light ray shown in the figure) by the same angle. That is, in the figure above, angle a is equal to angle b.

The ct' axis coincides with the worldline of O', which is inclined to the ct axis by the angle a. The worldline of the light ray (moving in the positive x and x' directions) lies exactly halfway between the ct and x axes. That is, the ct and x coordinates of any point on the light ray are equal, since, as we saw earlier, the light ray is characterized by $x - ct = 0$. From the invariance of the speed of light, the light ray must also lie exactly halfway between the ct' and x' axes as well, since $x' - ct' = 0$ is the equation of the light ray in the primed frame. Therefore, the ct' and x' coordinates of any point on the light ray must also be equal. A little time spent looking at figure A4.1 should convince you that this is true if the ct', x' axes are oriented as shown in the figure, with angle a equal to angle b. Note in particular that this could not be true if the ct' and x' axes formed a right angle with one another.

For a more rigorous proof, one can use the Lorentz transformation equations.

$$x' = \frac{x - vt}{\sqrt{1 - \dfrac{v^2}{c^2}}}$$

$$t' = \frac{t - \dfrac{vx}{c^2}}{\sqrt{1 - \dfrac{v^2}{c^2}}}$$

< 234 >

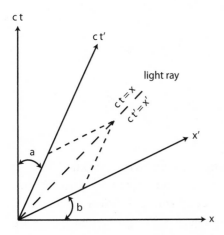

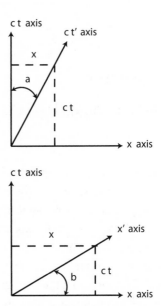

FIG. A4.1. A rotation in spacetime. Because of the geometry of spacetime (recall the presence of a minus sign in the spacetime interval), a coordinate transformation rotates both the time and space axes *inward*, that is, toward the path of the light ray in the figure.

FIG. A4.2. Rotation of the time axis and space axis, respectively.

Refer to figure A4.2. The ct' axis corresponds to the line $x' = 0$ (just like the ct axis corresponds to the line $x = 0$). Setting $x' = 0$ in the first of the Lorentz transformation equations above, we see that the equation of the ct' axis (i.e., the worldline of observer O') in the unprimed frame is just $x = vt$. Divide both sides of this equation by ct to get $x/(ct) = v/c$. Notice that the left-hand side of our last equation is just the slope of this line (i.e., the ct' axis), *measured with respect to the vertical ct axis.*

Similarly, the x' axis corresponds to the line $ct' = 0$ (just like the x axis corresponds to the line $ct = 0$). Setting $t' = 0$ in the second of the Lorentz transformation equations above, we get $t = vx/c^2$. Multiply both sides of this equation by c and divide both sides by x to get $ct/x = v/c$. The left-hand side of the last equation is the slope of the x' axis (i.e., the line $ct' = 0$) with respect to the horizontal x axis. Notice that the right-hand sides of our two slope equations are the same, namely, v/c. Therefore, since the slope of the ct' axis with respect to the ct axis and the slope that the x' axis makes with respect to the x axis are both equal to v/c, the angles a and b in the figure above must be equal.

Appendix 5
Time Dilation via Light Clocks

The previous derivation for time dilation, given in chapter 5, employed the Lorentz transformations. The following derivation essentially just uses the two principles of relativity and the Pythagorean theorem. Consider the following device, known as a "light clock": a rectangular box whose bottom and top are mirrored. In the bottom of the box is a flash gun that emits a photon (a particle of light) toward the top mirror. The photon, moving at speed c, travels the vertical distance d from the bottom to the top of the box, where it is reflected and returns to the mirror at the bottom, to begin the cycle all over again. The time interval for one round-trip of the photon will be regarded as one "tick" of this clock and is given by $2d/c$. (The above description applies to a frame in which the clock is at rest.)

Now consider a series of these light clocks, which are synchronized with one another in the earth frame, S(earth). At time $t = 0$, another light clock, which is at rest in S'(ship) and moving with speed v, relative to the earth frame, passes one of the clocks in S(earth) just at the moment when the clock in S'(ship) reads $t' = t = 0$. This situation is depicted in figure A5.1. We are interested in the time interval in each frame between the following two spacetime events: event 1, a photon is emitted from the bottom of the ship's light clock; and event 2, the photon is subsequently received at the bottom of the same light clock. Note that in S'(ship), the time between these two events can be measured by a *single* clock, because in S'(ship) both the emission and reception of the photon occur at the *same place*. In our example above, t' is therefore the proper time. We wish to calculate the time interval between these two events as measured in S'(ship) and S(earth). The observer in S'(ship) will see the photon go up and back, returning after a time $t_1' = 2d/c$. What corresponding time, t, will the observers in S(earth) measure?

For the observers in S(earth), the emission and reception of the photon occur in *different* places, because the light clock in S'(ship) is moving relative to observers in S(earth). Therefore the time (call it t_1) between the two events

< 236 >

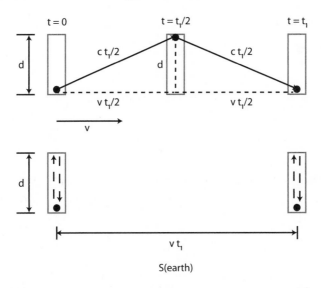

FIG. A5.1. Time dilation with light clocks. One "tick" of a light clock on the ship as measured by synchronized light clocks on the earth.

cannot be measured by a single clock in S(earth), but is deduced from the measurements of two clocks that are separated and synchronized in S(earth). Observers in both frames must agree that the following 3 events occur: the photon is emitted from a flash gun near the bottom mirror of the ship's clock, it strikes the top mirror of the same clock and is reflected, and the photon is received at the bottom mirror of the ship's clock. Also recall that the principles of relativity require that observers in each frame measure the speed of light to be c. From figure A5.1, we see that observers in S(earth) must therefore see the photon travel along a diagonal path. They will see this because the mirror is no longer in its original position, but has moved a distance $vt_1/2$ to the right, according to the S(earth) observers, by the time the photon reaches the top mirror. So in order for observers in both frames to see the photon hit the top mirror, it must travel along the diagonal path in S(earth). The (square of the) distance that the photon travels along the diagonal path to reach the top mirror at time $t = t_1/2$, is simply given by the Pythagorean theorem:

$$(ct_1/2)^2 = (vt_1/2)^2 + d^2.$$

Now, solve this for d to get

$$d = \frac{t_1}{2}\sqrt{c^2 - v^2}.$$

Pull out a factor of c from under the square root to obtain

$$d = \frac{ct_1}{2}\sqrt{1 - \frac{v^2}{c^2}}.$$

(The analysis is the same for the second half of the photon's trip.) Recall that the time for one tick as measured in S′(ship) was given by $t_1' = 2d/c$. Rewriting this in terms of d gives

$$d = \frac{ct_1'}{2}.$$

If we set the right-hand sides of these two formulas for d equal to one another, and cancel out the factors of $\frac{c}{2}$, we get

$$t_1' = t_1\sqrt{1 - \frac{v^2}{c^2}}.$$

This is the formula for time dilation that we found previously.

Note that our result is a direct consequence of the fact that observers in both frames must measure the speed of light to be c. Since the light travels a *greater distance* in S(earth), but at the *same speed* as in S′(ship), the round-trip time measured in S(earth) must be *longer* than that measured in S′(ship).

By the first principle of relativity, either set of observers is entitled to say that they are at rest and the observers in the other frame are the "moving" ones. If we are the observers in S′(ship), then we would consider a series of clocks at rest and synchronized in S′(ship) and a single light clock in S(earth) that is moving with velocity $-v$, relative to S′(ship). Now we would see the light clock in S(earth) as the one in which the photon makes the longer, diagonal trip. Hence, we would conclude that it is the clocks of S(earth) that are running slow, compared to our clocks.

You might be tempted to say that this effect is just hocus-pocus, the result of some peculiarity of the light clock we assumed in our discussion. However, we know this is not true, since the derivation using the Lorentz transformations made no assumption about a specific kind of clock being used and provided the same predictions. In fact, the time dilation effect of special relativity must apply to all clocks, regardless of construction. Let's see what would happen if that were not the case. Suppose we have a set of clocks of different types in an inertial frame, and we very gently and gradually accelerate them all up to some common constant velocity. If, subsequently, the clocks no longer tick at the same rate relative to each other, then we could establish the inertial frame in which their ticking rates do all agree as a "special" inertial frame, which is absolutely at rest. This would violate the principle of relativity, which says that all inertial frames are equivalent.

Length Contraction

Another frequently discussed consequence of special relativity is that not only do moving clocks run slow but moving meter sticks contract. This will not be of direct relevance for us, but for the sake of completeness—and since it can be obtained very easily from time dilation—we'll discuss this phenomenon of "length contraction" briefly.

Consider a stick at rest in S(earth) whose length as measured in that frame is L. (The length of an object measured in a frame in which the object is at rest is called its "proper length." As in the case of proper time, the term "proper" here does not mean "true" or "correct.") Let the left end of the stick be located at $x = 0$ and its right end at $x = L$. The light clock in S'(ship) travels to right with speed v as seen in S(earth). Let the time [as measured in S(earth)] at which the center of the clock passes the left end of the stick be $t = 0$ and the time when it passes the right end of the stick be $t = t_1$. So the length of the stick in S(earth) can be written as $L = vt_1$, that is, as measured in S(earth) the clock travels a distance L in a time t_1.

Now let's consider the situation from S'(ship). In this frame, the stick moves to the left at speed v. The left end of the stick passes the center of the clock in S'(ship) at $t' = t = 0$. The right end passes at $t' = t'_1$. Note that the two events, that is, the left and right ends of the stick passing the center of the clock in S'(ship), occur at the same place in S'(ship). Hence, the time between these events can be measured by a single clock in S'(ship), and so the time t' is the proper time. The length of the stick as measured in S'(ship) is then $L' = vt'_1$. Using the time dilation formula, $t_1' = t_1\sqrt{1 - \dfrac{v^2}{c^2}}$, we can write this as $L' = vt_1\sqrt{1 - \dfrac{v^2}{c^2}}$. But we know from our previous discussion that vt_1 is just equal to L, so we have

$$L' = L\sqrt{1 - \frac{v^2}{c^2}}.$$

This is the phenomenon of "length contraction," which is that the length of an object, as measured in a frame where the object is moving, is shorter than its length as measured in a frame where it is at rest, by a factor of $\sqrt{1 - \frac{v^2}{c^2}}$. As with time dilation, the effect is symmetrical in that observers in each frame will say that it is the other observer's stick that is shorter, compared to his own.[1]

1. One can show that lengths *perpendicular* to the direction of motion are unaffected.

Appendix 6
Hawking's Theorem

In this appendix, we will discuss a famous theorem by Stephen Hawking regarding time travel. The theorem appeared in his chronology protection conjecture paper (1992), but it holds *independently* of whether chronology protection is true or not. It is reasonable to assume that even an arbitrarily advanced civilization will be able to warp spacetime only in a *finite* region, in order to build a time machine. Hawking proved that, given certain assumptions, in order to build a time machine in a finite region of spacetime, one needs matter that violates the null energy condition, that is, negative energy.[1]

Here, we give a very rough sketch of his proof. Hawking uses the method of proof by contradiction. The techniques he applies are known as "global techniques" in relativity. These methods allowed Roger Penrose and Hawking to derive the famous "singularity theorems" in the 1960s and early 1970s. One advantage of using these techniques is that Hawking does not have to assume anything specific about the exact type of mass/energy used to build the time machine or the details of its construction. This makes his result very general and powerful.

Refer to figure A6.1 for the following discussion. Consider the spacelike surface S, which you can think of as a "snapshot" of space at one instant of time. (We assume S to be infinite in extent, so we can only show a finite portion of it). If we examine the point p, which lies to the future of S, we see that *every* past-directed timelike or lightlike curve from p (such as the dotted curve shown) intersects, or "registers" on, S. Therefore, what is going to happen at p can be predicted from information given on S (recall our light cone discussion from chapter 4), that is, from the shaded region of S where the past light cone of p intersects S. All such points p that have this property lie in what is called the (future) "domain of dependence of S," the region in the diagram called $D^+(S)$.[2] This is the region of spacetime (to the future of the surface S) that is

---

1. Hawking's analysis generalizes earlier work done by Frank Tipler in the 1970s.

2. To see this, pick any point in the region of the diagram, $D^+(S)$. Call it r. Draw the past light cone of r until the cone eventually intersects S, *keeping in mind that S extends infinitely far*. All past-

< 241 >

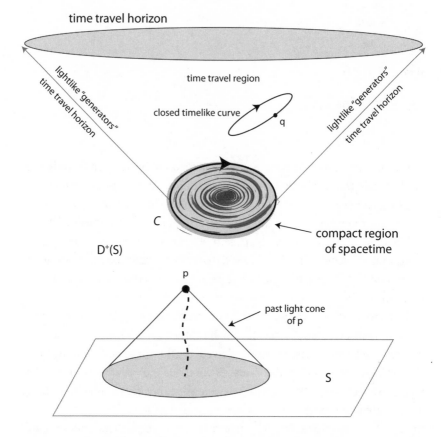

FIG. A6.1. A compactly generated time travel horizon. The lightlike generators of the horizon have no endpoints, but wind around and around in the compact region *C*.

predictable from information given on S. If we examine the point q, we see that it does not have this property, since there is a closed timelike curve through q. The past-directed part of this curve does not intersect S, and so does not "register" on S. Therefore, the point q does not lie in the domain of dependence, D⁺(S), of S. In other words, what is going to happen at q cannot be predicted by information given on S, since there is at least one curve (with a past-directed part) through q that does not "register" on S. The boundary of the region be-

---

directed timelike or lightlike curves from the point r must lie inside or on its past light cone, so if its past light cone intersects S, all the past-directed timelike or lightlike curves from r must also intersect S. Therefore, what is going to happen at r can be predicted from information on S.)

tween all points like q and points like p is what we have called the "time travel horizon" in figure A6.1. This boundary separates the region of spacetime containing closed timelike curves from the region that does not.

In his proof, Hawking first shows that the time travel horizon is made up of pieces of lightlike geodesics, called "generators." The generators are defined in such a way that there is only one generator passing through each point of the time travel horizon. As an analogy, think of the surface of a cylinder. Draw a series of lines on the cylinder that are parallel to the cylinder axis. We can think of these lines as "generators" of the cylinder in that, as we follow them along, they "trace out" the cylinder, and there is only one generator passing through each point.

In the case of the time travel horizon, it can be shown that no two points of the horizon can be connected by a timelike curve. Related to this is another important—and *nonobvious*—feature of the generators of the time travel horizon: they can have no past endpoints.[3] Endpoints are where generators enter or leave the horizon.

As an analogy, consider two light rays in a light cone in flat spacetime, which cross at the origin O. This is shown in the diagram on the left in figure A6.2. As a result of their crossing, the two points on the upper and lower parts of the light cone, labeled a and b, respectively, can be connected by a timelike curve (the dotted line). Similarly, if two nearby generators in the time travel horizon could cross one other at a point e, as shown, for example, on the right in figure A6.2, then the points labeled c and d could also be connected by a timelike curve. But then the two generators could not remain in the horizon, because no two points of the horizon can be connected by a timelike curve.

Since the generators of the time travel horizon have no past endpoints, what can they do? Well, why can't they just "stop"? In spacetime when a timelike or lightlike curve "just stops," that means it is simply not possible to extend it any farther. Such a curve indicates the presence of a spacetime singularity where space and time come to an end. For example, in the case of an observer following a timelike curve, this would mean that at some finite value of his wristwatch time, his existence suddenly comes to an end. He's "run out" of spacetime, as

---

3. *Technical note:* More precisely, the generators of the time travel horizon either have no past endpoints or past endpoints on the edge of S. But we have assumed in our argument that S is infinite in extent. Therefore, in our case, S has no edge. Also, although the generators can have no past endpoints, they can have future endpoints. But that does not affect our discussion.

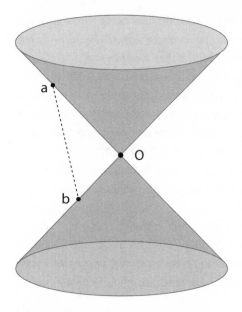

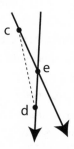

two nearby generators
in the time travel horizon with
a crossing point

two points on opposite sides of
the light cone in flat spacetime

FIG. A6.2. Crossing points of generators in spacetime.

it were. Since we can't know or control what comes out of a singularity, we want to exclude that possibility from our time machine construction.

If the generators of the time travel horizon don't run into a singularity, they could go off to infinity. That is, as we follow them backward into the past, they could get farther and farther away from the region of spacetime where we are building our time machine. But then information coming from extremely far away could affect the construction of our machine. Even the most advanced civilization can only manipulate spacetime in a *finite* region.

Therefore, in his proof, Hawking needs to precisely capture the notion of what it means to "build a time machine in a finite region of spacetime." To this end, he wants to, quite reasonably, exclude information coming in from a singularity or from infinity. OK, so if the generators don't run into a singularity or go off to infinity, what's left for them to do? Hawking takes as his definition of "building in a finite region" that the time travel horizon is "compactly generated." This means that if we follow the generators of the time travel horizon toward the past direction, they enter and remain within some bounded (more precisely, "compact") region of spacetime C, spiraling around and around, and never leave the time travel horizon (this is illustrated in figure A6.1).

Now consider two such generators that lie very close to each other in the time travel horizon. Follow these generators backward, in the past direction. Since they are entering a *bounded* region of spacetime, instead of, say, going off to infinity, they must start to converge. (As we follow the generators past-ward, the region of spacetime they border gets smaller, so they must converge toward the past.) Hawking also initially assumes that the null energy condition holds. (Actually, the weaker, averaged null energy condition is sufficient.) Recall that this is similar to the weak energy condition, but along light rays. This condition guarantees that light rays are always *focused*—never defocused—by gravity. If the generators in the time travel horizon start to converge, and if the null energy condition holds, one can show that the light rays must cross eventually each other, as depicted on the right in figure A6.2. (It can be shown that this will occur within a finite distance along the rays.)

But once the light rays cross, they leave the time travel horizon at the point where they cross. This means that the crossing point is a past endpoint, which lies on the time travel horizon. But this is a contradiction, because, as Hawking showed earlier, the generators of the time travel horizon *have* no past endpoints. The only way these generators can start to converge but never cross, and thus not have past endpoints, is if the null energy condition is violated. Thus, in order to build a time machine in a finite region of spacetime, given Hawking's assumptions, one requires negative energy. In closing, we again emphasize that this conclusion *does not depend* on the validity of the chronology protection conjecture.

Caveats: Some Objections to Hawking's Arguments

Not everyone agrees with Hawking's criterion for "building a time machine in a finite region of spacetime," that is, his condition that time travel horizons should be compactly generated. One person who disagrees is Amos Ori, at the Technion in Israel. He feels that Hawking's condition is too restrictive a condition for "buildability." Ori has published a time machine model that initially consists of vacuum plus "dust" (i.e., noninteracting particles). Specifically, there is *no negative energy* used in his construction. Nevertheless, closed timelike curves eventually develop in the model. This is because the time travel horizon in Ori's model is not compactly generated, so it evades Hawking's theorem. However, this implies that Ori's time machine model will contain either naked singularities (not hidden inside of black holes) or "internal infinities," which we will describe below.

We can divide Ori's objections into two parts, which we might call the "finiteness" argument and the "causal control" argument. The finiteness argument relates to the requirement of building a time machine in a finite region. Ori argues that we have to be careful about what we mean by "finite region." For example, do we mean a finite region of three-dimensional *space* or a finite region of four-dimensional *spacetime*? Ori's criterion is that the time machine should be "compactly constructed," that is, that the time machine originate in an *initially* finite region of three-dimensional *space*. His time machine model does have this property. Ori argues that if this region of space is initially finite, one has control over it at the time when the time machine is turned on. However, if the time travel horizon is not compactly generated, in the sense of Hawking, it is possible that as a result of forming the time machine, naked singularities may develop, or that this region might be "blown up" (enlarged) by the subsequent evolution of the spacetime, to form what is called an "internal infinity."

One way of thinking about the latter is to take a point from spacetime, and imagine "moving" that point until it is "infinitely far away" (in some technically appropriate sense). If the worldline of an observer approaches this point, it will take the observer an infinite proper time to reach it. It's a little like running toward a receding goalpost, which always moves away too fast for you to ever reach it. So instead of a singularity, we have created a "point at infinity." Ori remarks that such internal infinities arise in typical black hole models. Therefore, he argues that if such an infinite region forms from the initially finite region of three-dimensional space upon construction of our time machine, this is not necessarily cause for concern. It simply indicates that this just happens to be the way the spacetime will evolve according to the laws of general relativity. Since we can collapse matter in an initially finite region of space to form black holes, we shouldn't worry about internal infinities forming as a result making a time machine. We comment, however, that in the black hole cases, the internal infinities are hidden behind event horizons. In the time machine models they would not be (or at least not need to be). To us, this seems to be a crucial difference.

Ori points out that in black hole spacetimes, the black hole, although filling only a finite region of three-dimensional space in the external world (i.e., its horizon is of finite size at any moment of time), has an infinite *four-dimensional*

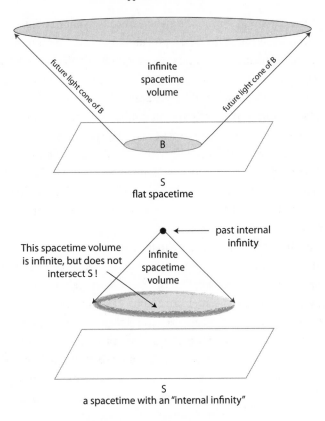

FIG. A6.3. A spacetime with a (past) "internal infinity."

interior volume. This is because the horizon exists forever (neglecting the Hawking black hole evaporation process).

We point out that the same is true for the future of any bounded region of a spacelike surface in flat spacetime. It's just that in that case, there is no horizon for it to hide behind. For example, in the top diagram in figure A6.3, take S to be a spacelike surface in *flat* spacetime (with no time machines, black holes, etc.). The circular region labeled B (which would be a sphere in real three-dimensional space, which S is supposed to represent), is a bounded region of the spacelike surface S. Draw the future light cone of that region. It will contain an infinite volume of spacetime because the cone expands forever, getting larger as time goes on.

Now let us examine a possible problem with internal infinities. Suppose that such an infinite region arises "smack in the middle of the spacetime," as it were, where you are building your time machine. Ken Olum points out that if Ori's construction leads to the development of past "internal infinities," there are likely to be problems. In this case, by past internal infinities we mean: not hidden behind event horizons, arising in the region where we are trying to construct our time machine, and whose past light cone opens out to infinity. We would be creating a place that has an infinite spacetime volume within its past light cone that does not intersect the initial spacelike surface S, so that the past internal infinity does not lie within the domain of dependence of S. The initial conditions within this past light cone can affect the formation of the time machine, but they are not determined by the initial conditions on the surface S, over which we have control, which seems very odd. This is illustrated in the bottom diagram in figure A6.3.

A second, more fundamental issue that Ori raises is the "causal control" argument, that is, Hawking's use of compactly generated time travel horizons as an argument for causal control of the region of spacetime exterior to the region of closed timelike curves. The attempt to control a region in which closed timelike curves develop can be difficult, even in principle. If we go back to our diagram in figure A6.1, we see that the region of spacetime that contains closed timelike curves is not in the domain of dependence of the surface S, that is, it lies outside the region marked $D^+(S)$, by definition. The domain of dependence is defined to be the region of spacetime that can be predicted by information given on S. The boundary of this region is the time travel horizon. Therefore the region of closed timelike curves, which lies inside the time travel horizon, lies outside of $D^+(S)$. This means that whatever happens in that region cannot be predicted or controlled from the initial information given on S. So, Ori argues, we don't know for sure whether closed timelike curves will appear there or not, or whether, if they do, they will form in just the ways we expect (e.g., with no singularities or no internal infinities). Put the other way, a compactly generated time travel horizon does not guarantee that you will get closed timelike curves, that is, a time machine. So even the definition of what it means to "build" a time machine is a bit of a tricky business.[4]

Ori mentions that the above discussion implies that causality arguments alone cannot determine whether closed timelike curves will form. However, he also points out that if one assumes, in addition, that the spacetime is "smooth,"

---

4. Some of these concerns have also been raised by the philosophers of science: John Earman, Christopher Smeenk, and Christian Wüthrich (see http://philsci-archive.pitt.edu/archive/00004240/01/TimeMachPhilSciArchive.pdf).

that is, has no abrupt jumps in spacetime structure across the time travel horizon, then one can use causality plus smoothness to conclude that closed timelike curves must occur in his model and in a number of others. So Ori views causality plus smoothness as providing a kind of "limited causal control" over whether closed timelike curves will develop as a result of one's manipulations. He emphasizes that this property of limited causal control is the best that one can hope to achieve in any time machine model, independently of whether the time travel horizon is compactly generated or not. Therefore, Ori feels that Hawking's use of compactly generated time travel horizons as an argument for insuring causal control over the region up to and including the time travel horizon is not conclusive.

Ori is only mildly concerned if time machine construction involves naked singularities or internal infinities that we cannot control. For example, he points out that we are living comfortably with a naked singularity in our past, namely, the big bang in which our universe began. It does not seem to have affected our ability to predict the outcomes of experiments in our laboratories.

Since the time travel horizon in Ori's model is not compactly generated, this means that either naked singularities or internal infinities (or both) must be present in his model. If the generators of the time travel horizon, when traced back into the past, do not spiral around in a compact region, they must either end in a singularity or at a point at infinity. Our point of view, in contrast to Ori's, is in agreement with that of Hawking, namely, that one should avoid naked singularities and internal infinities in scenarios designed to produce time machines. We would say that it's bad enough if you may not be able to uniquely predict what happens beyond a time travel horizon. We feel that things are made worse by naked singularities or regions at infinity that you also cannot control. Also, the occurrence of one naked singularity at the beginning of time (i.e., the big bang) worries us less than what happens to our powers of predictability if we manufacture naked singularities every time we manipulate matter in a some appropriate way. The possibility of naked singularities (or internal infinities) popping up all over the place, as it were, is a lot more troubling to us.

Whether one likes or dislikes naked singularities and internal infinities depends to some extent on personal taste, and on what one is willing to accept. The consensus of the relativity community, we think, favors Hawking's view, as do we. However, we should remind the reader (and ourselves!) that in science the majority view is not always necessarily the correct one. Ultimately what counts are *nature*'s preferences, and she has not yet shown us her cards.

Appendix 7

Light Pipe in the Mallett Time Machine

Consider a section of the helical light pipe, with length l, which is short enough to be considered straight. Let the laser power, the energy per second, flowing perpendicularly through the circular left face of the pipe be denoted as

$$P = \frac{E}{t},$$

where E is the energy and t is the time. The radius of the pipe is r, so the cross-sectional area of the pipe is $A = \pi r^2$, and the volume of this section of pipe is $V = \pi r^2 l$. The energy density, the energy per unit volume, is then

$$\frac{E}{V} = \frac{E}{\pi r^2 l} = \frac{Pt}{\pi r^2 l}.$$

Let us choose l to be the distance the light beam travels in a time t = 1sec, so that $l = ct = c \times 1\text{sec}$. (Don't worry that this is a very large distance; we could have picked any time, since it will cancel out in the next step.) Substituting this into the equation above, we get

$$\frac{E}{V} = \frac{E}{\pi r^2 c(1\text{sec})} = \frac{P(1\text{sec})}{\pi r^2 c(1\text{sec})} = \frac{P}{\pi r^2 c}.$$

Define the energy per unit length along the pipe, as it winds around the z axis, as $\varepsilon = \frac{E}{l}$. Then using the equation just above, we get that

$$\varepsilon = \frac{E}{l} = \frac{E}{\pi r^2 l}\left(\pi r^2\right) = \frac{E}{V}\left(\pi r^2\right) = \frac{P}{c}.$$

So we have that the energy per unit length along the pipe is

$$\varepsilon = \frac{P}{c}.$$

Using Einstein's mass-energy relation, $\varepsilon = mc^2$, we can write the mass per unit length, m, along the pipe as

< 250 >

$$m = \frac{P}{c^3}.$$

To convert m, the mass per unit length along the laser beam as it winds *around* the z axis, to the total mass per unit length *along the z axis* in the circulating laser beam, let us refer back to figure 13.2. Consider a (finite) length tightly wound helical light pipe, so that each winding sits right on top of the previous winding with no spacing between them, with length L (along the z axis), and radius R_0. There is one winding of the light pipe per each $2\pi R_0$. If we call the total number of windings N, then $N = \frac{L}{d} = \frac{L}{2r}$, where d is the diameter of the light pipe and r is its radius. On the other hand, the *total* length of the light pipe, L_1, as measured *around* the z axis is

$$L_1 = N \times 2\pi R_0 = \frac{\pi R_0 L}{r}.$$

The mass per unit length as measured around the z axis is

$$m = \frac{M}{L_1},$$

where M is the total mass equivalent of the laser energy in the entire light pipe.

To convert this to mass per unit length as measured along the z axis, m', we have

$$m' = \frac{M}{L} = \left(\frac{M}{L_1}\right)\left(\frac{L_1}{L}\right) = m\left(\frac{L_1}{L}\right).$$

Now, using our expression for L_1 above and our result that $m = \frac{P}{c^3}$, we get

$$m' = m\left(\frac{L_1}{L}\right) = \frac{P}{c^3}\left(\frac{\pi R_0 L}{rL}\right)$$

or

$$m' = m\left(\frac{\pi R_0}{r}\right) = \frac{P}{c^3}\left(\frac{\pi R_0}{r}\right).$$

With our chosen values of $r = 1$ millimeter $= 10^{-3}$ m, and $R_0 = 0.5$m, with $\pi \approx 3.14$, this gives $\frac{\pi R_0}{r} \approx 10^3$, as stated in the text.

Bibliography

Alcubierre, M. "The Warp Drive: Hyper-Fast Travel within General Relativity." *Classical and. Quantum Gravity* 11 (1948): L73–L77.

Antippa, A., and A. Everett. "Tachyons, Causality and Rotational Invariance." *Physical Review D* 8 (1973): 2352–60.

Barcelo, C., and M. Visser. "Scalar Fields, Energy Conditions, and Traversable Wormholes." *Classical and Quantum Gravity* 17 (2000): 3843.

———. "Traversable Wormholes from Massless Conformally Coupled Scalar Fields." *Physics Letters B* 466 (1999): 127–34.

Baxter, S. *The Time Ships.* New York: HarperCollins Publishers, 1995.

Benford, G. *Timescape.* New York: Pocket Books, 1981.

Benford, G., D. Book, and W. Newcomb. "The Tachyonic Antitelephone." *Physical Review D* 2 (1970): 263.

Bilaniuk, O., N. Deshpande, and E. Sudarshan. "Meta Relativity." *American Journal of Physics* 30 (1962): 718.

Borde, A. "Geodesic Focusing, Energy Conditions and Singularities." *Classical and Quantum Gravity* 4 (1987): 343–56.

Brown, L., and G. Maclay. "Vacuum Stress between Conducting Plates: An Image Solution." *Physical Review* 184 (1969): 1272.

Casimir, H. "On the Attraction between Two Perfectly Conducting Plates." *Proceedings of the Koninklijke Nederlandse Akademie Van Wetenschappen B* 51 (1948): 793–95.

Clark, C., B. Hiscock, and S. Larson. "Null Geodesics in the Alcubierre Warp Drive Spacetime: The View from the Bridge." *Classical and Quantum Gravity* 16 (1999): 3965.

Clee, M. *Branch Point.* New York: Ace Books, 1996.

Davies, P. *About Time: Einstein's Unfinished Revolution.* New York: Simon and Schuster, 1995. See esp. chap. 10, "Backwards in Time," 219–32, and chap. 11, "Time Travel: Fact or Fantasy?" 233–51.

Davies, P., and S. Fulling. "Radiation from Moving Mirrors and from Black Holes." *Proceedings of the Royal Society of London Series A* 356 (1977): 237–57.

Deutsch, D. "Quantum Mechanics Near Closed Timelike Lines." *Physical Review D* 44 (1991): 3197–217.

< 253 >

Deutsch, D., and M. Lockwood. "The Quantum Physics of Time Travel." *Scientific American*, March 1994, 68–74.

Dummett, M. "Causal Loops." In *The Nature of Time*, edited by R. Flood and M. Lockwood, 135–69 Oxford: Basil Blackwell Ltd., 1986.

Einstein, A. Appendix I, in *Relativity: The Special and the General Theory*. New York: Crown Publishers 1962, 115–20.

Einstein, A., and N. Rosen. "The Particle Problem in the General Theory of Relativity." *Physical Review* 48 (1935): 73–77.

Everett, A. "Warp Drive and Causality." *Physical Review D* 53 (1996): 7365.

———. "Time Travel Paradoxes, Path Integrals, and the Many Worlds Interpretation of Quantum Mechanics." *Physical Review D* 69 (2004): 124023.

Everett, A., and T. Roman. "A Superluminal Subway: The Krasnikov Tube." *Physical Review D* 56 (1997): 2100.

Everett, H. "Relative State Formulation of Quantum Mechanics." *Reviews of Modern Physics* 29 (1957): 454–62.

Feinberg, G. "Particles that Go Faster than Light." *Scientific American*, February 1970, 69–77.

———. "Possibility of Faster-Than-Light Particles." *Physical Review* 159 (1967): 1089.

Fewster, C. "A General Worldline Quantum Inequality." *Classical and Quantum Gravity* 17 (2000): 1897–911.

Fewster, C., and S. Eveson. "Bounds on Negative Energy Densities in Flat Space-Time." *Physical Review D* 58 (1998): 104016.

Fewster, C., K. Olum, and M. Pfenning. "Averaged Null Energy Condition in Space-times with Boundaries." *Physical Review D* 75 (2007): 025007.

Fewster, C., and L. Osterbrink. "Averaged Energy Inequalities for the Non-Minimally Coupled Classical Scalar Field." *Physical Review D* 74 (2006): 044021.

———. "Quantum Energy Inequalities for the Non-Minimally Coupled Scalar Field." *Journal of Physics A* 41 (2008): 025402.

Fewster, C., and T. Roman. "On Wormholes with Arbitrarily Small Quantities of Exotic Matter." *Physical Review D* 72 (2005): 044023.

Finazzi, S., S. Liberati, and C. Barcelo. "On the Impossibility of Superluminal Travel: The Warp Drive Lesson." Second prize of the 2009 FQXi essay contest "What is Ultimately Possible in Physics?" http://xxx.lanl.gov/abs/1001.4960.

———. "Semiclassical Instability of Dynamical Warp Drives." *Physical Review D* 79 (2009): 124017.

Ford, L. "Constraints on Negative Energy Fluxes." *Physical Review D* 43 (1991): 3972.

——— "Quantum Coherence Effects and the Second Law of Thermodynamics." *Proceedings of the Royal Society of London A* 364 (1978): 227–36.

Ford, L., and T. Roman. "Averaged Energy Conditions and Quantum Inequalities." *Physical Review D* 51 (1995): 4277.

———. "Negative Energy, Wormholes, and Warp Drive." *Scientific American*, January 2000, 46–53.

———. "Quantum Field Theory Constrains Traversable Wormhole Geometries." *Physical Review D* 53 (1996): 5496–507.

Friedman, J., and A. Higuchi. "Topological Censorship and Chronology Protection." *Annalen Der Physik* 15 (2006): 109–28.

Friedman, J., K. Schleich, and D. Witt. "Topological Censorship." *Physical Review Letters* 71 (1993): 1486–489; erratum, *Physical Review Letters* 75 (1995): 1872.

Frolov, V., and I. Novikov. "Physical Effects in Wormholes and Time Machines." *Physical Review D* 48 (1993): 1057–65.

Fuller, R., and J. Wheeler. "Causality and Multiply Connected Space-Time." *Physical Review* 128 (1962): 919.

Galloway, G. "Some Results on the Occurrence of Compact Minimal Submanifolds." *Manuscripta Mathematica* 35 (1981): 209–19.

Gao, S., and R. Wald. "Theorems on Gravitational Time Delay and Related Issues." *Classical and Quantum Gravity* 17 (2000): 4999–5008.

Gödel, K. "An Example of a New Type of Cosmological Solution of Einstein's Field Equations of Gravitation." *Reviews of Modern Physics* 21 (1949): 447–50.

Gott, J. "Closed Timelike Curves Produced by Pairs of Moving Cosmic Strings: Exact Solutions." *Physical Review Letters* 66 (1991): 1126–29.

Hartle, J. *Gravity: An Introduction to Einstein's General Relativity.* San Francisco: Addison Wesley, 2003.

Hawking, S. "Chronology Protection Conjecture." *Physical Review D* 46 (1992): 603–11.

———. "Particle Creation by Black Holes." *Communications in Mathematical Physics* 43 (1975): 199–220; erratum, *Communications in Mathematical Physics* 46 (1976): 206.

———. "The Quantum Mechanics of Black Holes." *Scientific American*, January 1977, 34–40.

Heinlein, R. "By His Bootstraps." In *The Menace from Earth*. Riverdale, NY: Baen Publishing Enterprises, 1987.

———. *The Door into Summer*. New York: The New American Library, 1957.

Hiscock, B. "Quantum Effects in the Alcubierre Warp Drive Spacetime." *Classical and Quantum Gravity* 14 (1997): L183–88.

Kay, B., M. Radzikowski, and R. Wald. "Quantum Field Theory on Spacetimes with a Compactly Generated Cauchy Horizon." *Communications in Mathematical Physics* 183 (1997): 533–56.

Kim, S., and K. Thorne. "Do Vacuum Fluctuations Prevent the Creation of Closed Timelike Curves?" *Physical Review D* 43 (1991): 3929–47.

Krasnikov, S. "Hyperfast Interstellar Travel in General Relativity." *Physical Review D* 57 (1998): 4760.

Kruskal, M. "Maximal Extension of Schwarzschild Metric." *Physical Review* 119 (1960): 1743.

Lobo, F. "Exotic Solutions in General Relativity: Traversable Wormholes and 'Warp Drive' Spacetimes." *Classical and Quantum Gravity Research* 1–78 (2008).

Lobo, F., and M. Visser. "Fundamental Limitations on 'Warp Drive' Spacetimes." *Classical and Quantum Gravity* 21 (2004): 5871.

Lossev, A., and I. Novikov. "The Jinn of the Time Machine: Nontrivial Selfconsistent Solutions." *Classical and Quantum Gravity* 9 (1992): 2309–21.

Mallett, R. "The Gravitational Field of a Circulating Light Beam." *Foundations of Physics* 33 (2003): 1307–14.

Mallett, R., and B. Henderson. *Time Traveler: A Scientist's Personal Mission to Make Time Travel a Reality.* New York: Basic Books, 2006.

Morris, M., and K. Thorne. "Wormholes in Spacetime and Their Use for Interstellar Travel: A Tool for Teaching General Relativity." *American Journal of Physics* 56 (1988): 395–412.

Morris, M., K. Thorne, and U. Yurtsever. "Wormholes, Time Machines, and the Weak Energy Condition." *Physical Review Letters* 61 (1988): 1146–49.

Natário, J. "Warp Drive with Zero Expansion." *Classical and Quantum Gravity* 19 (2002): 1157–66.

Novikov, I. *The River of Time.* Cambridge: Cambridge University Press, 1998.

Olum, K. "Geodesics in the Static Mallett Spacetime." *Physical Review D* 81 (2010): 127501.

———. "Superluminal Travel Requires Negative Energies." *Physical Review Letters* 81 (1998): 3567.

Olum, K., and A. Everett. "Can a Circulating Light Beam Produce a Time Machine?" *Foundations of Physics Letters* 18 (2005): 379–85.

Olum, K., and N. Graham. "Static Negative Energies Near a Domain Wall." *Physics Letters B* 554 (2003): 175–79.

Ori, A. "Formation of Closed Timelike Curves in a Composite Vacuum/Dust Asymptotically-Flat Spacetime." *Physical Review D* 76 (2007): 044002.

Parker, L. "Faster-Than-Light Intertial Frames and Tachyons." *Physical Review* 188 (1969): 2287.

Pfenning, M., and L. Ford. "The Unphysical Nature of 'Warp Drive.'" *Classical and Quantum Gravity* 14 (1997): 1743.

Rolnick, W. "Implications of Causality for Faster-Than-Light Matter." *Physical Review* 183 (1969): 1105.

Roman, T. "On the 'Averaged Weak Energy Condition' and Penrose's Singularity Theorem." *Physical Review D* 37 (1988): 546–48.

———. "Quantum Stress-Energy Tensors and the Weak Energy Condition." *Physical Review D* 33 (1986): 3526–33.

Sagan, C. *Contact.* New York: Simon and Schuster 1985.

Slusher, R., L. Hollberg, B. Yurke, J. Mertz, and J. Valley. "Observation of Squeezed States Generated by Four-Wave Mixing in an Optical Cavity." *Physical Review Letters* 55 (1985): 2409.

Taylor, B., B. Hiscock, and P. Anderson. "Stress-Energy of a Quantized Scalar Field in Static Wormhole Spacetimes." *Physical Review D* 55 (1997): 6116.

Taylor, E., and J. Wheeler. *Spacetime Physics: Introduction to Special Relativity*. 2nd ed. New York: W. H. Freeman and Company, 1992.

Tipler, F. "Energy Conditions and Spacetime Singularities." *Physical Review D* 17 (1978): 2521–28.

———. "Rotating Cylinders and the Possibility of Global Causality Violation." *Physical Review D* 9 (1974): 2203–6.

———. "Singularities and Causality Violation." *Annals of Physics* 108 (1977): 1–36.

Thorne, K. *Black Holes and Time Warps: Einstein's Outrageous Legacy*. New York: W. W. Norton, 1994.

Toomey, D. *The New Time Travelers*. New York: W. W. Norton, 2007.

Urban, D., and K. Olum. "Averaged Null Energy Condition Violation in a Conformally Flat Spacetime." *Physical Review D* 81 (2010): 024039.

———. "Spacetime Averaged Null Energy Condition." *Physical Review D* 81 (2010): 024039.

Van Den Broeck, C. "A 'Warp Drive' with More Reasonable Total Energy Requirements." *Classical and Quantum Gravity* 16 (1999): 3973.

van Stockum, W. "Gravitational Field of a Distribution of Particles Rotating about an Axis of Symmetry." *Proceedings of the Royal Society of. Edinburgh* 57 (1937): 135–54.

Vilenkin, A. "Gravitational Field of Vacuum Domain Walls and Strings." *Physical Review D* 23 (1981): 852.

Visser, M. *Lorentzian Wormholes: From Einstein to Hawking*. Woodbury, NY: American Institute of Physics Press, 1995.

———. "The Quantum Physics of Chronology Protection." Contribution to *The Future of Theoretical Physics and Cosmology*, a conference in honor of Professor Stephen Hawking on the occasion of his 60th birthday, edited by G. Gibbons, E. Shellard, and S. Rankin, 161–73. Cambridge: Cambridge University Press, 2003.

———. "The Reliability Horizon for Semi-Classical Quantum Gravity: Metric Fluctuations Are Often More Important than Back-Reaction." *Physics Letters B* 415 (1997): 8–14.

———. "Traversable Wormholes: Some Simple Examples." *Physical Review D* 39 (1989): 3182–84.

———. "Traversable Wormholes: The Roman Ring." *Physical Review D* 55 (1997): 5212.

Visser, M., S. Kar, and N. Dadhich. "Traversable Wormholes with Arbitrarily Small Energy Condition Violations." *Physical Review Letters* 90 (2003): 201102.

Wells, H. G. *The Time Machine*. New York: Tor Books, 1992.

Index

Page numbers in italics refer to figures.

Aeschylus, 136

aether, 25–30

air table, 18

Alcubierre, Miguel, 6–7, 117–21

Alcubierre warp drive spacetime, 6–7, 117–21, *118*, 159, 185–86, *187*

Al-Khalili, Jim, *Black Holes, Wormholes, and Time Machines*, 184n2

Ampère, André-Marie, 24

Anderson, Paul, 184

Anderson, Poul, *Tau Zero*, 57n2

Andromeda galaxy, distance to, 1

angular momentum, conservation of, 70

antimatter, 159

antiparticle, 41, 159

Antippa, Adel, 68–69, 70

"arrow of time" concept, 76–88; additional arrows, 87n7; causal arrow, 84–87; cosmological arrow, 87–88; thermodynamic arrow, 81–84

atomic clocks, 12, 17, 36, 59, 99, 104

Augustine, Saint, 10

averaged energy conditions, 167–69. *See also specific types*

averaged null energy condition, 168–69, 177–78, 179–80, 245

averaged weak energy condition, 168, 169, 172, 175, 176, 179–80

"back-reaction" effect, 121, 191, 192–93

"banana peel mechanism" idea, 7–8, 144–45, 154, 157, 220

Barcelo, Carlos, 121, 179, 185

baryon number, conservation of, 71–72

Baxter, Stephen, *The Time Ships*, 145

Benford, Gregory, 67; *Timescape*, 67n1

Berra, Yogi, 22, 49

Bilaniuk, O., 64, 66

black holes: evaporation effect, 164–65, 167; event horizon and, 106, 109, 206; explanation of, 6, 106; forward time travel and, 6; internal infinities and, 246–249; singularities inside of, 159; time dilation effect near, 106–11; as time machines, 109–11; wormholes similar to, 114

Book, D. L., 67

bootstrap paradoxes. *See* information paradoxes

Borde, Arvind, 168

Bose-Einstein condensates, 221

Carroll, Lewis, 218

Casimir, Hendrik, 164

Casimir effect, 164, 167, 169, 172, 175–78, 180, 182–83, 220

Cauchy horizon, 128

causal loops: closed, 130, 134; inconsistent, 4, 131–32, 142–43; self-consistent, 141

causality, principle of, 84–87, 130–32, 248–49

cause-and-effect: coincidence and, 84–85; light cone and, 42, 46, 47–48, 84–87; time travel and, 6, 248–49

< 259 >

CERN. *See* European Organization for
Nuclear Research (CERN)
Chaplin, Charlie, 105
charge-to-mass ratios, 90–91
Chew, Geoffrey, 63
chronology horizon, 128, 190
chronology protection conjecture, 8, 144,
189–95, 198, 199, 222, 241, 245
Clark, Chad, 121
Clee, Mona, *Branch Point*, 145
clever spacecraft scenario, 137–40
clocks: atomic, 12, 17, 36, 59, 99, 104;
biological, 60–61; gravity and, 99–101;
light, 52, 236–40; special relativity and,
49–52; synchronization and simultaneity, 36–39, 49; time measurement by, 5,
6, 12, 16–17, 32–33, 54–58
closed timelike curves: cylinder time
machines and, 198–209; cylindrical
universe and, 196–97; explanation of,
4; forbidden by chronology protection conjecture, 194; Gödel's universe
and, 198; Gott's model, 213–14; in
Hawking's theorem, 242, 242–43, 245,
248–49; time travel horizon and, 193; in
wormhole time machines, 128
Coleridge, Samuel Taylor, 112
Colton, Charles Caleb, 112
consistency paradoxes, 53, 136, 140–44.
See also grandfather paradox
constant velocity, 17
cosmic strings, 8–9, 184, 209–13,
216–17
cosmic string time machine, 209, 213–17,
214, 222–23
cosmological constant, 189
Coulomb, Charles-Augustin de, 24
cryogenic sleep, 60–61
"curved spacetime" idea: general theory
of relativity and, 89, 101–3, 102, 108;
gravity as result of, 6; local flatness of,
181–83, 219–20
cylindrical universe, 196–97; properties of,
197; topology of, 196–97

Dahich, Naresh, 185
dark energy, 189, 220
Davies, Paul, *How to Build a Time Machine*,
173–74
Davis, E. W., 223n2
decoherence phenomenon, 152–53, 155
density matrices, 149n6
Deshpande, N., 64, 66
designer spacetimes, 115
Deutsch, David, 8, 149, 151–57
Dickens, Charles, 49
Dirac field, 173, 180
directions in space, laws of physics and,
69–71, 76–77
directions in time, laws of physics and,
77, 84
distances in space, 1, 219, 221
domain walls, 212
Dummett, Michael, 15
dust, 198n1
Dylan, Bob, 42, 158

Earman, John, 248n4
Eddington, Arthur Stanley, 76, 105
Einstein, Albert: derivation of Lorentz
transformations, 227, 231; energy-
mass relation equation, 7, 13, 31,
40–41, 95, 209; field equations, 102–3,
114–17, 200–1, 218; general theory of
relativity, 4, 6, 59, 85, 89–111, 158–59;
special theory of relativity, 2, 5–6,
30–31, 39–41, 42, 59; statement about
general theory of relativity, 115–16;
thought experiments, 92–95; wormhole
idea, 114
electromagnetic field, quantum inequalities in, 170–73, 180
electromagnetic radiation energy, 41
electromagnetism, 24–25, 30: gravity *vs.*,
89–91; infinitely long systems studied
in, 199
electron, behavior of, 146–47, 149–50,
152
Eliot, T. S., 42

energy, conservation of: Einstein's equation and, 31, 39–41; law of, 3, 79, 134–35, 143–44, 162–63, 194; tachyons and, 71–74; Wells's time machine's violation of, 13–14

energy density in squeezed vacuum state, 166, 166–67, 172

energy-time uncertainty principle, 86n5, 163, 169, 172. *See also* uncertainty principle

entanglement phenomenon, 85–86n5

entropy, 79–84, 87–88, 202

equilibrium state, 80, 82

equivalence, principle of: curved spacetime idea and, 219–20; gravity's effect on clocks, 99–101; mass and, 91–94, 93, 94, 95, 105; negative mass and, 160

European Organization for Nuclear Research (CERN), Geneva, 5

event horizon, 106–7, 109, 114, 116, 206, 246, 248

Everett, Allen, 9, 62–63, 70–71, 75, 201–2; "Time Travel Paradoxes, Path Integrals, and the Many Worlds Interpretation of Quantum Mechanics," 141–43, 155

Everett, Hugh, 8, 146, 147, 150–53

Eveson, Simon, 173

exotic matter (negative energy), 158–80; antimatter contrasted with, 159; averaged energy conditions and, 167–69; Casimir effect, 164, 167, 169, 172, 175–78, 180, 182–83; classical fields and, 178–80; dark energy as, 220; Hawking's theorem, 7, 181, 198–202, 205, 208, 217, 241–59; Krasnikov tube's use of, 122–24, 186; negative energy as, 158–67; negative mass contrasted with, 159–63, 188; physical restrictions on, 7; quantum inequalities and, 169–73, 180; quantum interest and, 175–78, 179; quantum mechanics laws regarding, 163–67; warp bubbles' use of, 119, 186; wormholes' use of, 116–17, 183–85, 222–23

Faraday, Michael, 24, 89–90

Farscape, 6

Feinberg, Gerald, 63, 64

Fermi National Laboratory, 5

Fewster, Chris, 169n2, 173, 177, 178, 179–80, 188

Feynman, Richard, 223

Finazzi, Stefano, 121

first law of thermodynamics. *See* energy, conservation of

Ford, Larry, 7, 75, 148, 169–73, 183, 185–86

Foster, Jodie, 114

Foundations of Physics Letters, 201

frame of reference, 17–18. *See also* inertial frame of reference; noninertial frame of reference

Friedman, John, 168–69

Frolov, Valery, 129

Fuller, Robert, 114

Fulling, Stephen, 173–74

Galilean transformations, 19–22, 26–33; velocity transformations derivation, 225–26

Galilei, Galileo, 19, 92

Galloway, Greg, 168

Gao, Sijie, 124

Geller, Uri, 223n2

general theory of relativity, 89–111; bending of light by the sun test, 104–5; black holes and, 106–11; "classical" tests of, 103–11; "curved spacetime" idea in, 89, 101–3, 180, 181–83; discovery of, 4; Einstein's views on, 115–16; explanation of, 101–3; gravitational redshift test, 103–4; gravity and light, 94–95; precession of the perihelion test, 103; principle of equivalence, 91–94, 105, 219–20; space-time structure in, 6; tidal forces, 95–99; time dilation effect and, 59, 236–40

geodesics, 101–2, 111, 178, 207–8, 243

Geroch, Bob, 190

Gilbert, W. S., 136
Gilliam, Terry, 144
global inertial frame of reference, 98–99
Gödel, Kurt, 198, 201
Gold, Thomas, 76
Gott, Richard, 209, 213–17
Graham, Noah, 176–77, 180, 183,
    188–89
grandfather paradox, 4, 7; billiard ball
    version, 129–35, 131, 145, 151–52; as
    consistency paradox, 53, 136; many
    worlds interpretation solution, 149,
    150–51; self-consistent solution to,
    132–34, 133, 141–44
granularity of space, 184–85
Gravel, Pierre, 187
gravitational mass, 91–92, 160
gravitational redshift, 103–4, 106
gravity: of black holes, 106–11; earth's
    gravitational field, 100; Einstein's
    theory of, 4, 6; electromagnetism vs.,
    89–91; light and, 94–95; Newton's law
    of, 89, 90–92, 97–98, 160–61; principle
    of equivalence, 92–94, 105; theory of
    semiclassical, 190–91, 192; tidal flex-
    ing, 95–99; time and, 99–101; used for
    time dilation, 129
ground state of energy, 180
Guth, Alan, The Inflationary Universe, 211

half-life, 59–60, 71, 77n1
Hawking radiation, 164–65
Hawking, Stephen: chronology protec-
    tion conjecture, 8, 144, 191–95, 198,
    199, 222, 241, 245; evaporation effect
    predicted by, 164–65; exotic matter
    theorem, 7, 181, 198–202, 205, 208,
    217, 241–49; singularity theorems, 159,
    243, 243–44
Heinlein, Robert: By His Bootstraps, 144;
    The Door into Summer, 60–61
Heisenberg, Werner, 163, 169
Hiscock, Bill, 121, 184
Hubble, Edwin, 189

inconsistent causal loops, 4, 131–32,
    142–43
inertial frame of reference: definition of,
    18; earth as, 26, 54; explanation of,
    18–21, 20; Galilean transformations,
    19–21; gravity and, 95–99; principle
    of general covariance, 95; principles
    of relativity and, 30–31, 54, 74, 89, 96;
    speed of light and, 30, 33, 64–66; sun's
    center of, 26; tachyons and, 69
inertial mass, 91, 160
information paradoxes, 136–40, 151–52
initial conditions, 16, 78, 82, 132–33, 134,
    248
interference phenomenon, 23, 27–29, 28
interferometer, 27
internal infinity, 246–49, 247
invariant interval, 33–36, 44, 46, 54–58;
    derivation, 232–33; lightlike, 45, 48;
    spacelike, 45, 48, 126; timelike, 45,
    48, 54

jinnee balls scenario, 137–40

Kar, Sayan, 185
Kay, Bernard, 191–92
Kibble, Tom, 209, 212
Kim, Sun-Won, 190–91
Krasnikov, Serguei, 7, 119–21, 122
Krasnikov tubes, 7, 122–24, 123, 126,
    186–87, 194
Kruskal, Martin, 114

Lamb, Charles, 22
Lamb, Willis, 164
Lamb shift, 164
Large Hadron Collider (LHC), 5
Larson, Shane, 121
Lawrence Berkeley National Laboratory,
    62, 63
length contraction phenomenon, 239–40
Liberati, Stefano, 121
Light: barrier, 2, 4, 6, 39–40, 44–46,
    64; bending of by sun, 104–5; early

research on, 22–23; gravity and, 94–95; Maxwell's research on, 24–25; photons, 40; Young's research on, 23–24. *See also* speed of light; superluminal travel

light clocks, time dilation via, 52, 236–40, 237

light cones, 42–48, 47, 109, 110; causality and, 47–48, 84–87, 247–49; curved spacetimes represented by, 108–10; Krasnikov tube and, 122–24, 123; Lorentz transformations and, 46–47; significance of, 44; special relativity and, 42; time travel horizon, 127–29, 243–44, 244; warp bubbles and, 117–18, 121

light pipes in Mallett time machine, 9, 201–2, 203, 207–9, 250–51

light-year, length of, 1

liquid helium, 60, 221

Lobo, Francisco, 187–88

local inertial frame of reference, 98

Lorentz transformations, 31–41; clock synchronization and simultaneity, 36–39, 49, 51–52, 236, 237–39; clocks in, 32–33; derivation, 227–31; equations, 31–33, 32, 37–38, 39, 65; invariant interval and, 33–36, 215, 232–33; inverse equations, 52n1; light barrier and, 39–40, 63–64; light cone and, 46–47; orientation to coordinate axes in x,t plane, 35–36, 124, 234–35; reinterpretation principle and, 66–68, 73; superluminal reference frames and, 68, 70–71

Lossev, A., "The Jinn of the Time Machine," 137–40

macrostate of a system, 80–82, 150

Mallett, Ronald: rotating cylindrical time machine model, 8–9, 200–209, 250–51; *Time Traveler*, 9, 200, 207

Mandelstam, Stanley, 63

many worlds interpretation of quantum mechanics, 8, 146–54; decoherence phenomenon in, 152–53, 155; grand

father paradox and, 150–51; information paradox resolutions in, 151–52; slicing and dicing problem, 155–57, 220; time machine and, 148–49

Marchildon, Louis, 70

massless particles, 40–41

mass spectrometer, 90–91

mathematician's proof paradox, 136–37, 151

Maxwell, James Clerk, 24–25, 89–91

Maxwell's equations, 25, 26, 30, 76, 89–90

Mercer, Johnny, 158

Michelson, Albert, 25–30

Michelson-Morley experiment, 25–30, 28, 31, 38, 233

microstate of a system, 80–81, 88, 150

Milky Way galaxy, distances across, 1

Miller, P. Schuyler, "As Never Was," 141n4

momentum, conservation of, 39–41, 70, 162, 187–88, 231

Morley, Edward, 25–30

Morris, Mike S., 115–17, 126–34, 178, 183; "Wormholes, Time Machines, and the Weak Energy Condition" (with Thorne and Yurtsever), 4, 177

Mossbauer effect, 104

motion, first law of, 18

motion, second law of, 160–62

muons, 60

naked singularities, 206–7, 246, 248–49

Natário, José, 117, 121, 187

negative energy. *See* exotic matter

negative mass, 75, 159–63, 188

neutron stars, 105–6, 107, 129

Newcomb, W. A., 67

Newton, Isaac: law of conservation of momentum, 187–88; law of gravitation, 89, 90–92, 97–98, 160–61, 179; laws of motion, 18, 77–79, 134, 160–62; spacetime theories, 42

Newton, Roger B., 86n6

noninertial frame of reference: clocks at rest in, 99–101; explanation of, 18; principle of general covariance, 95

nonminimally coupled scalar field (NMCSF), 178–80, 185, 189

Novikov, Igor, 129, 134, 143; "The Jinn of the Time Machine," 137–40

Olum, Ken, 9, 124, 176–77, 178, 180, 183, 188–89, 201–2, 206, 207–8, 221, 248

Ori, Amos, 245–49

Osterbrink, Lutz, 179–80, 188

paradoxes: in backward time travel, 4, 7–8, 15–16, 52–58, 129–35, 136–45; consistency, 53, 136, 140–44; grandfather, 4, 7, 53, 130–34, 136, 141–44, 149, 150–51; information, 136–40, 151–52; mathematician's proof, 136–37; twin, 5, 52–58, 119, 126–29; types of, 136

parallel worlds, 8, 145–57. *See also* many worlds interpretation of quantum mechanics

Parker, Leonard, 68

particles, formula for energy of, 40–41

Penrose, Roger, 82n3, 159, 241

perihelion shift, 103

perpetual motion machines, 194

Petty, Tom, 89

Pfenning, Mitch, 177, 178, 185–86

photons, 40, 236–38

Physical Review, 8, 64, 149, 151, 207–8

Physical Review Letters, 4, 216–17

Planck, Max, 163

Planck length, 183, 184, 185–86, 187, 210

Planck's constant, 163, 169–70

Plante, Jean-Luc, 187

post hoc ergo propter hoc fallacy, 85

Pound, R. V., 104

precession of the perihelion, 103

preferred frame of reference for speed of light, 25–30

"proper time," 52, 54–58, 127–28, 197, 208, 236, 239, 246

proton decay, 72–74

Proxima Centauri, distance to, 1

Puthoff, H. E., 223n2

Pythagorean theorem, 28, 34, 236–40

quantum gravity, 109, 222

quantum inequalities: constraints of, 181–83, 201, 218, 219–20, 222–23; Davies-Fulling analysis, 173–75; explanation of, 169–73, 180; Fewster-Eveson derivation of, 173; in flat spacetime, 181–82; physical interpretation of, 170; possibilities for circumventing, 188–89; quantum interest effect, 175–78, 179; sampling function of, 182–83; warp drives and, 185–86; wormholes and, 183–85

quantum interest effect, 175–78, 179

quantum mechanics: Copenhagen interpretation, 147, 148–49; energy-time uncertainty principle, 86n5, 163, 169, 172; entanglement phenomenon, 85–86n5; light's particle-like properties discovered through, 24, 40; many worlds interpretation, 8, 146–54; negative energy described by, 163–67; theory of, 146–47; time-reversal invariance and, 77n1, 134; uncertainty principle, 86n5, 165–66, 192

quantum physics: black hole evaporation effect, 164–65; examples of negative energy in, 164–67; squeezed states of light, 165–66; squeezed vacuum state, 166–67

quantum stress-energy tensor, 191–92

Radzikowski, Marek, 191–92

Rebka, G. A., 104

redshift, gravitational, 103–4, 106–7

reinterpretation principle, 66–68, 73

reliability horizon, 192, 193, 193

"rest" energy, 40–41, 66, 210

rest frame, 38–39, 65–66, 72–74, 182

Reviews of Modern Physics, 8, 146

Riemann, Bernhard, 181–83
Riemann curvature tensor, 181–83
"ring of wormholes" spacetime, 192–93, 222
Rolnick, W. B., 67
Roman, Thomas, 7, 75
Rosen, Nathan, 114
rotating cylinder time machines, 8–9, 198–209

Sagan, Carl, 195, 223; *Contact*, 114
Schleich, Kristen, 168–69
Schwarzschild, Karl, 103
Schwarzschild radius, 107, 108
Schwarzschild solution, 103, 206
science fiction: development of, 5–6; parallel universe idea in, 145, 146, 151, 157; reactionless drives in, 187; superluminal travel in, 1, 2, 117, 159, 221; time travel in, 2–4, 7–8, 11–16, 60–61, 65, 138, 144; wormholes in, 6–7, 144. *See also specific authors and titles*
second law of thermodynamics, 79–84; chronology protection and, 194; cosmological arrow of time and, 87–88; Eddington on, 76; explanation of, 80–84; self-existing objects and, 138; violations of, 16, 82, 167, 169, 179, 223n2
semiclassical gravity, theory of, 190–91, 192
Shakespeare, William, 218
Shellard, Paul, 212
singularity: big bang, 82n3; black hole, 109; naked, 206–9, 246, 249; spacetime, 243–44; theorems, 159, 200n2, 241
slicing and dicing problem, 155–57, 220
Sliders, 6
Smeenk, Christopher, 248n4
Smith, Calvin, 169n2
solar eclipse, light-bending effect during, 104–5
Somewhere in Time (film), 138

sound barrier, 2
space, measurements of, 16–21
space, perception of, 14–16
space-time structure: "curved spacetime" idea, 89, 101–3, 102, 112–14, 117–18, 180, 181–83, 219–20; designer spacetimes, 115; in general relativity theory, 6; internal infinity in, 246–49, 247; invariant interval in, 33–36, 54–58, 232–33; light cone idea, 42–48; spaghettification, 107–8
special theory of relativity: absolute and relative concepts, 42–47; acceleration of matter through speed of light prohibited by, 63–64; backward time travel and, 6; clocks and, 49–52; curved spacetime and, 182; first principle, 30–31; forward time travel and, 4–5, 58–60; inertial frames of reference in, 96–97; "massless" particles and, 40–41, 59–60; second principle, 30; superluminal travel as violation of, 2, 5–6, 39–40, 90, 117; tachyon travel and, 66–68; time dilation effect and, 221, 236–40; twin paradox and, 52–54, 119, 126–29, 127
spectral lines, 164
speed of light: early research on, 22–23; Lorentz transformations and, 33; Maxwell's discovery of, 24–25; Michelson-Morley experiment, 26–30; reference frame for, 25; second principle of relativity and, 30, 37–38, 90; value of, 1; warp bubble circumvention of, 117–18
speed of sound, 2, 25–26
squeezed states of light, 165, 165–66
squeezed vacuum state, 166, 166–67
Stargate SG1, 6
Star Trek, 117, 181, 223; antimatter in, 159; time travel depicted in, 2
Star Trek Deep Space Nine, 6
Steinbeck, John, 10
Stern-Gerlach apparatus, 146–47

Sticklin, Marylee, 62–63

string theory, 75, 188, 215, 222

subatomic particles, forward time travel of, 5, 59–60

Sudarshan, E. C. G., 64, 66, 67

superluminal particles. See tachyons

superluminal reference frames, 68–71

superluminal travel; backward, 124–35, 125, 194–95; chronology protection and, 194; concept of, 2; Einstein's special theory of relativity and, 2, 5–6, 39–40, 63–64; Krasnikov tube and, 7, 122–24, 185–86; practical problems, 221–23; of tachyons, 63–66, 215; twin paradox and, 53–54, 126–29, 127; through warp bubbles, 7, 117–21; through wormholes, 112–17, 113–14

supernova explosions, 105–6

tachyons, 63–68; decline of interest in, 74–75; directionality problem, 69–71; experimental evidence for, 71–75; explanation of, 63–64; paradoxes and, 64–66; science fiction depiction, 67n1; string theory and, 75, 215; superluminal reference frames and, 68–71, 126

Targ, Russell, 223n2

Taylor, Brett, 184

Taylor, E., Spacetime Physics (with Wheeler), 47–48

test fields, 178

thermodynamics, laws of. See energy, conservation of; second law of thermodynamics

Thorne, Kip S.: Black Holes and Time Warps, 114, 127; vacuum fluctuation studies, 189–91; "Wormholes, Time Machines, and the Weak Energy Condition" (with Morris and Yurtsever), 4, 177; wormholes investigated by, 114–17, 126–34, 142, 183, 185

tidal forces, 95–99, 96, 97, 107–8, 116, 119, 182n1, 183

time, measurement of, 16–21: Galilean transformations, 19–21; gravity and, 99–101. See also clocks

time, perception of: differences from spatial perception, 11–12; external vs. internal time, 12–13; subjectivity of, 10–11

time dilation effect, 50; avoidance in warp bubbles, 119; black hole orbit contrasted with, 6; explanation of, 5, 49–52; gravitational, 129; impracticality of, 59; via light clocks, 236–40; for muons, 60; near black holes, 106–7, 111; possibility of travel, 221; ways of achieving, 53

"time gate" concept, 2

time machines: black holes used as, 109–11; chronology protection and, 194; concept of, 3–4; cosmic string, 209, 213–17, 214, 222–23; destruction and chronology protection, 189–95; exotic matter needed for, 7, 181, 208–9; Krasnikov tube used as, 124, 126, 186–87; Mallett's model, 8–9, 200–9, 250–51; many worlds theory tested with, 150–51; Ori model, 245–49; restrictions on, 218; in science fiction, 11–16, 53; warp bubbles used as, 7, 117–21, 122–24, 126, 185–86; Wells's depiction of, 11–16, 53, 61, 135; wormholes used as, 7, 112–17, 124–29, 137–44

time travel: consistency problems in, 14–16, 132–34, 136, 140–44; meaning of, 5; practical problems, 221–23; "rate of travel" notion, 12; science fiction depictions of, 2–4; speculations on, 9, 218–23; superluminal travel and, 2–3; through wormholes, 13–14; Wells's depiction of, 11–16. See also directional headings

time travel, backward, 3–4; "banana peel mechanism" idea, 7–8, 144–45, 154, 157, 220; billiard ball experiment, 129–35, 145, 151–52; common-sense objections to, 6; conservation of energy and, 134–35; paradoxes in, 4, 7–8, 15–16, 52–58, 129–35, 136–45; parallel worlds idea, 8, 145–54; rotating cylinder idea, 8–9, 198, 199–209; slicing and dicing problem, 155–57; superluminal, 124–35, 194–95; tachyons and, 64–68

time travel, forward, 4–7, 49–61; black holes and, 6; cryogenic sleep and, 60–61; energy requirements for, 5, 58–59; practical considerations and experiments, 58–60; reality of, 4–5; of subatomic particles, 5, 59–60; superluminality and, 5–6; tachyons and, 63–68; warp bubbles and, 117–21; wormholes and, 112–17

time travel horizon, 127–29, 190–93, 193, 201, 242, 243–46, 247–49

time-reversal invariance, 77–79

Tipler, Frank, 168, 200, 204, 241n1

topological censorship theorem, 168–69

topological defects, 212–13

triangles, in curved space, 182

12 *Monkeys* (film), 144

Twilight Zone, "No Time Like the Past" episode, 144

twin paradox, 52–57; explanation of, 52–54; invariant interval and proper time, 54–58, 55, 58; muon experiment, 60; reference frames and, 54; time dilation effect and, 5, 53; wormhole time machine and, 126–29, 127

uncertainty principle, 86n5, 165–66, 192. *See also* energy-time uncertainty principle

universe, expansion of, 87–88, 185, 189, 210–11, 220

Urban, Doug, 178, 221–22

vacuum fluctuations, 163–64, 166, 190–94

vacuum solutions, 103

Van Den Broeck, Chris, 186–87

van Stockum, W. J., 199–200, 201, 204, 206

Vilenkin, Alex, 211–12; *Cosmic Strings and Other Topological Defects*, 212

virtual particles in space. *See* vacuum fluctuations

Visser, Matt, 179, 183, 184, 185, 187–88, 191–92, 222–23; *Lorentzian Wormholes*, 117

Wald, Bob, 124, 190, 191–92

warp bubbles, 117–21; "back-reaction" effect, 121; disadvantages of, 7, 119–21; exotic matter used in, 119, 185–86; explanation of, 6–7; impossibility of steering, 119–21, 120; Natário model, 117, 187; as time machines, 7, 117–21, 122–24, 126, 185–86

warp drives, 194; Alcubierre model, 6–7, 117–21, 159, 185–86, 187; Krasnikov tube, 7, 122–24, 126, 186–87; quantum inequalities and, 170, 180, 185–86, 219–20; reactionless, 187–88; restrictions on, 218; scientific studies, 4; on *Star Trek*, 2; types of, 6–7; Van Den Broeck model, 186–87

weak energy condition: cosmic strings and, 184; dark energy and, 189; exotic matter and, 158–59, 167; explanation of, 4, 115–16; null energy conditions compared with, 245; for test fields, 178; violations of, 187, 200n2. *See also* averaged weak energy condition

Wells, H. G., *The Time Machine*, 11–16, 53, 61, 135, 145

Wheeler, John, 102, 114, 184–85; *Spacetime Physics* (with Taylor), 47–48

white dwarf stars, 105, 106

Witt, Don, 168–69

wormholes, 113; Barcelo-Visser, 179, 185; concept of, 112–17; conservation of energy in, 134–35; cubical, 117, 184, 222–23; current knowledge about, 7; double-hole time machine model, 124–26; event horizon of, 114; exotic matter used in, 116–17, 184–85; grandfather

wormholes (continued)
paradox and, 130–34, 151–52; length scale describing, 183–84; NMCSF used in constructing, 179, 185, 189; nontraversable, 114; quantum inequalities and, 170, 180, 183–85, 219–20; restrictions on, 218; science fiction depictions of, 6; scientific investigation of, 6–7; single-hole time machine model, 126–29; size of, 183–84; superluminal travel through, 113–14; as time machines, 7, 112–17, 124–29, 137–44; traversable, 114–17, 168–69, 177–78; Visser-Kar-Dahich (VKD), 185; Visser's ring of, 192–93, 222

Wüthrich, Christian, 248n4

Young, Thomas, 23–24
Yurtsever, Ulvi, 126–34; "Wormholes, Time Machines, and the Weak Energy Condition" (with Morris and Thorne), 4, 177